£3.00

ical

Technology 2:

Science and Calculations

also in this series

Electrical Installation 1:
Theory and Regulations
Maurice Lewis

Electrical Installation Technology 2:
Science and Calculations

Maurice L. Lewis

B Ed (Hons), FIElecIE

Head of Electrical Section, Faculty of the Built Environment
at Luton College of Higher Education

Stanley Thornes (Publishers) Ltd

First published in 1988 by Hutchinson Education
Reprinted 1989

Reprinted in 1990 by:
Stanley Thornes (Publishers) Ltd
Old Station Drive
Leckhampton
CHELTENHAM GL53 0DN
England

ISBN 0 7487 0378 0

Printed and bound in Great Britain at
Scotprint Ltd, Musselburgh

Contents

Preface

Electrical students often find great difficulty in understanding the electrotechnology side of their industry, since it involves learning many well-founded principles and related formulae which are not altogether easily grasped. This book has been written with the intention of making students' studies a little easier and it attempts to cover some of the important topic areas found in electrical courses at City & Guilds Part I and Part II level. It will be of immense value to students studying courses 185, 201, 232 and 236. Electrical installation students will find that the book covers a large number of topics in their science course scheme and it supplements the 'Theory and Regulations' book written by the same author.

Briefly, this book consists of seven chapters, starting with terminology as an introduction to important terms, formulae and the mathematical know-how to perform calculations. Chapter 2 concerns basic circuit theory which covers most of the established electrical principles such as Ohm's Law and Kirchhoff's Law. Chapter 3 expands some of the previous knowledge and deals with alternating current circuits involving circuit connections, phasor diagrams and power-factor problems. Chapter 4 is extremely important to Part II students and deals with electrical machines and their operating principles. Transformers and rectifiers are covered in Chapter 5, while heating and lighting calculations are covered in Chapter 6. Chapter 7 concerns instrument connections.

The final chapter contains past examination papers and multiple-choice questions for students to do as helpful revision. Answers to these questions and chapter exercise questions will be found at the end of the book.

M.L.L.
1988

Acknowledgements

The author wishes to thank several manufacturers
and organizations who have contributed information
to make this book possible, particularly:

Chloride Industrial Batteries Ltd
Compact Instruments Ltd
Institution of Electrical Engineers
The City and Guilds of London Institute—for
allowing their past examination questions to be used.
All solutions are those of the author.

The author also wishes to thank his wife for the
patience she has shown during the many hours of
writing this book.

CHAPTER ONE

Terminology

After reading this chapter you will be able to:

1 State the meaning of a number of electrotechnical terms and units.

2 Use a number of metric prefixes and their symbols.

3 Apply a number of transposition of formulae techniques to solve simple problems.

4 Apply Pythagoras's theorem and trigonometrical ratios to solve right-angled triangle problems.

5 Solve simple area problems of quadrilaterals and circles.

6 Draw simple straight-line and curved graphs.

7 Solve problems associated with ratios, proportion and percentages.

It was mentioned in 'Installation Theory and Regulations' (by the same author) that students studying electrical courses found considerable difficulty in grasping basic terminology used in their industry. To some extent, this is also true for the terminology associated with electrical principles and science using electrotechnical terms. For example, one often comes across a statement written by a student which reads: 'The voltage flowed through the circuit', when really it should have read: 'The *current* flowed through the circuit'. Omission and incorrect use of symbols is another area of concern and a student will quite unwittingly write down a formula which commences: $V/A = 45$. But 45 what? Such brief information is typical and totally unacceptable. The student is probably trying to find the resistance of something and of course should have started with 'R' (the term symbol for resistance). It should be noted that the term symbol for current is wrong and the formula is written as: $R = V/I$. The answer will finally be expressed in ohms using the Greek letter (Ω) meaning 'omega'. One wonders whether or not the use of calculators can be blamed for poor presentation, where errors often seem to occur and where essential working is found missing.

SI Units

The UK and many other European countries deal with the international system of units (known as SI units) which developed from the metric system. There are six basic units worth considering and these are explained in the following table.

Quantity	Unit name	Unit symbol
length	metre	m
mass	kilogram	kg
time	second	s
electric current	ampere	A
temperature	kelvin	K
luminous intensity	candela	cd

It should be noted that the kelvin is the unit of thermodynamic temperature but for practical purposes the degree Celsius scale is used. Both have identical intervals, i.e. 1 K = 1°C but the Kelvin scale begins at absolute zero, (the coldest temperature possible, −273°C). This means of course that freezing point (0°C) becomes 273 K and boiling point (100°C) becomes 373 K.

Definitions of derived SI quantities

From the basic SI Units come numerous derived units and those most commonly found in electrical engineering work are:

Capacitance: The property of a capacitor to store an electric charge. The farad or microfarad is the unit of capacitance.

Charge: The excess of positive or negative electricity on a body or in space, or passing at a point in an electric circuit during a given time. The coulomb is the unit of charge and is the quantity of electricity transported in one second by a current of one ampere.

Current: The flow or transport of negative charges along a path or around a circuit. The ampere is the unit of current.

Electromotive force: The force necessary to cause the movement of charges. The volt is a unit of electromotive force which will cause one ampere to flow through a resistance of one ohm.

Energy: The capacity for doing work. It is sometimes referred to as the ability of matter or radiation to do work. The joule is the unit of energy and is the work done when a force of one newton is displaced through a distance of one metre in the direction of the force.

Force: The cause of mechanical displacement or motion. It is measured in terms of the effects it produces. The newton is the unit of force.

Frequency: Of an alternating current waveform, it describes the number of repetitive cycles that occur in one second. The duration of one cycle is termed the periodic time. The hertz is the unit of frequency. In the UK the public supply frequency is 50 Hz.

Illuminance: The amount of light in lumens falling on a unit area of one square metre. The lux is the unit of illuminance.

Impedance: The ratio of voltage and current in r.m.s. terms for a.c. quantities. The ohm is the unit of impedance.

Inductance: The property of an inductor producing a magnetic field when carrying current. The henry is the unit of inductance.

Luminous flux: The capacity of radiant energy to produce light. It is sometimes thought of as the flow of light measured in lumens.

Magnetic flux: The phenomenon associated with invisible 'lines of force' in the neighbourhood of magnets and electric currents. The unit of magnetic flux is the weber.

Magnetic flux density: A measure of the quantity of magnetic flux spread over a given area. Its unit is the tesla.

Magnetizing force: The magnetomotive force per unit length of a magnetic circuit. Its unit is the ampere-turns per metre.

Magnetomotive force: The effects of setting up magnetic flux or the force required to cause magnetic flux to flow. The ampere is the unit of magnetomotive force.

Permeability: The ratio of magnetic flux density and the magnetizing force. The unit of permeability is the henry per metre.

Potential difference: The cause of movement of electric charge from one point to another, its unit is the volt.

Power: The dissipation of energy or the product of voltage and current. The unit of power is the watt.

Reactance: The property of inductors and capacitors to resist the flow of alternating current. The ohm is the unit of reactance.

Reluctance: The ratio of the magnetic force acting round a magnetic circuit to the resulting magnetic flux. Its unit is called the ampere per weber.

Resistance: The property of a resistor which resists the flow of charge through it. The ohm is the unit of resistance.

Resistivity: The resistance when measured between the opposite faces of a unit cube of given material. The unit of resistivity is called the ohm metre.

The above definitions are shown in the following table where it will be noticed that the quantity symbols are sloping (italic) and these are the ones that should be used by students when they are writing out formulae.

Quantity	Symbol	Unit name	Unit symbol
capacitance	C	farad	F
charge	Q	coulomb	C
current	I	ampere	A
electromotive	E	volt	V
energy	W	joule	J
force	F, f	newton	N
frequency	f	hertz	Hz
illuminance	E	lux	lx
impedance	Z	ohm	Ω
inductance (self)	L	henry	H
luminous flux	F	lumen	lm
magnetic flux	Φ (phi)	weber	Wb
magnetic flux density	B	tesla	T
magnetizing force	H	ampere per metre	A/m
magnetomotive force	F	ampere-turn	A, At
permeability	μ	henry per metre	H/m
potential difference	V	volt	V
power	P	watt	W
reactance	X	ohm	Ω
reluctance	S	ampere per weber	A/Wb
resistance	R	ohm	Ω
resistivity	ϱ	ohm-metre	Ω m

Metric system

One advantage of the metric system is the ease with which one can use prefix symbols to replace multiples and sub-multiples of various units. The following table shows some of the more common prefixes used in electrical work together with their meanings and multiplying factors.

Prefix	Symbol	Meaning		Multiplication factor
mega	M	1 000 000	(10^6)	one million times
kilo	k	1 000	(10^3)	one thousand times
hecto	h	100	(10^2)	one hundred times
deca	da	10	(10^1)	ten times
deci	d	0.1	(10^{-1})	one tenth of
centi	c	0.01	(10^{-2})	one hundredth of
milli	m	0.001	(10^{-3})	one thousandth of
micro	μ	0.000 001	(10^{-6})	one millionth of

Useful metric weights and measures

Linear measure

10 millimetres (mm)	=	1 centimetre (cm)	10 m	=	1 decametre (dam)
10 cm	=	1 decimetre (dm)	10 dam	=	1 hectometre (hm)
10 dm	=	1 metre (m)	10 hm	=	1 kilometre (km)

Square measure

$100 \text{ mm}^2 = 1 \text{ cm}^2$
$100 \text{ cm}^2 = 1 \text{ dm}^2$
$100 \text{ dm}^2 = 1 \text{ m}^2$
$100 \text{ m}^2 = 1 \text{ dam}^2$
$100 \text{ dam}^2 = 1 \text{ hm}^2$
$100 \text{ hm}^2 = 1 \text{ km}^2$

Cubic measure

$1000 \text{ mm}^3 = 1 \text{ cm}^3$
$1000 \text{ cm}^3 = 1 \text{ dm}^3$
$1000 \text{ dm}^3 = 1 \text{ m}^3$

Liquid measure

10 centilitres (cl) = 1 decilitre (dl)
10 dl = 1 litre (l)
10 l = 1 decalitre (dal)
10 dal = 1 hectolitre (hl)

Useful equivalents

1 litre of water = 1 kilogram (kg)
1000 kg = 1 tonne (t)
1000 l = 1 m³
1 unit of electricity = 1 kilowatt hour (kWh)
1 kWh = 3.6 megajoules (MJ)
1 joule (J) = 1 newton metre (Nm)
1 newton (N) = 0.102 kgf
Note: 1 kgf = 9.8 N

Examples using metric prefix symbols

1   Convert the following:
    a)   10 500 000 joules to megajoules
    b)   0.5 volts to millivolts
    c)   20 202 watts to kilowatts
    d)   600 milliamperes to amperes
    e)   0.000 003 farads to microfarads
Solution
    a)   1 000 000 J = 1 MJ
        10 500 000 J = 10.5 MJ
    b)   1 V = 1000 mV
        0.5 V = 0.5×1000 mV
        = 500 mV
    c)   1000 W = 1 kW
        20 202 W = 20.202 kW
    d)   1000 mA = 1 A
        600 mA = 600/1000
        = 0.6 A

    e)   1 F = 1 000 000 µF
        0.000 003 F = 0.000 003×1 000 000 µF
        = 3 µF

2   Express the following in more convenient units:
    a)   6600 watt hours
    b)   0.75 litres
    c)   778 000 ohms
    d)   0.099 milliamperes
    e)   50 000 kilograms
Solution
    a)   6.6 kWh
    b)   75 cl
    c)   778 kΩ
    d)   99 µA
    e)   50 t

Miscellaneous terms and symbols

Terms	Symbols	Often used to signify
Greek letters		
alpha	α	angle or temperature coefficient of resistance
eta	η	efficiency
mu	μ	micro
omega	Ω	ohm, resistance
phi	Φ, ϕ	magnetic flux, angle or phase difference
pi	π	circle constant— circumference/diameter
rho	ϱ	resistivity
sigma	Σ, σ	sum of, or conductivity
theta	θ	angle or temperature
Signs		
	$\approx$	approximately equal to
	$\propto$	proportional to
	∞	infinity
	$>$	greater than
	$<$	less than
	$\geq$	equal to or greater than
	$\leq$	equal to or less than

Useful electrical formulae

Quantity symbols		Notes
Q	=	It
V	=	IR
P	=	VI
W	=	Pt

$W = Fd$ where d is distance in metres

$R = \varrho l/a$ where a is cross-sectional area

$R_t = R_0(1 + \alpha t)$ where R_t is final resistance R_0 is resistance at 0°C t is change in temperature

$V = E + IR$ battery charging, where V is battery terminal volts E is battery e.m.f.

$V = E - IR$ battery discharging where R is battery internal resistance I is the charging or discharging current

$f = 1/T$ where T is the periodic time for an a.c. quantity f is the frequency

for a.c. sinusoidal systems where I_a is the average (mean) value of current I_r is the r.m.s. (effective) value of current V_a is the average (mean) value of voltage V_r is the r.m.s. (effective) value of voltage I_m is the maximum value of current V_m is the maximum value of voltage

$I_a = 0.637\, I_m$
$I_r = 0.707\, I_m$
$V_a = 0.637\, V_m$
$V_r = 0.707\, V_m$

$R = R_1 + R_2 + R_3 + \ldots$ for resistors in series
$L = L_1 + L_2 + L_3 + \ldots$ for inductors in series
$C = C_1 + C_2 + C_3 + \ldots$ for capacitors in parallel

$$R = \frac{1}{1/R_1 + 1/R_2 + 1/R_3 + \ldots}$$ for resistors in parallel

$$L = \frac{1}{1/L_1 + 1/L_2 + 1/L_3 + \ldots}$$ for inductors in parallel

$$C = \frac{1}{1/C_1 + 1/C_2 + 1/C_3 + \ldots}$$ for capacitors in series

$F = BIl$ for force on a conductor (F) where B is the flux density of the magnetic field I is the current flowing in the conductor l is the length of conductor in the field

$\dfrac{V_p}{V_s} = \dfrac{N_p}{N_s} = \dfrac{I_s}{I_p}$ for transformer ratios where V_p is the primary winding voltage N_p is the primary winding turns

I_p is the primary winding current V_s is the secondary winding voltage N_s is the secondary winding turns I_s is the secondary winding current

$V = V_s - V_L$ where V is the cable volt drop V_s is the supply volts V_L is the load terminal voltage

$V = l \times I \times mV/A/m$ where V is the volt drop $mV/A/m$ is the millivolt drop per ampere per metre I is the load current l is the length of run in metres

a.c. circuits

$V = IZ$
$P = VI \cos \phi$ for 1-phase a.c. circuits
$Z = \sqrt{R^2 + X^2}$
$P = \sqrt{3}\, V_L I_L \cos \phi$ for 3-phase a.c. balanced circuits*

$V = IX_L$ for a purely inductive circuit
$X_L = 2\pi fL$ where X_L is inductive reactance

$V = IX_C$ for purely capacitative
$X_C = 1/2\pi fC$ circuit where X_C is capacitative reactance

$\cos \phi = P/VI$ where $\cos \phi$ is the power factor of the circuit

Induction and magnetism

$e = Blv$ for induced e.m.f. in a conductor where B is the flux density of the magnetic field l is the length of conductor in the field v is the velocity of the conductor

*For 3-phase star connections the voltage to neutral (phase voltage) is $V_p = V_L/\sqrt{3}$ where V_L is the line volts. Also phase current = line current. For 3-phase delta connections the line volts are the phase volts, but the phase current $I_p = I_L/\sqrt{3}$.

$e = \Phi / t$ where e is the average induced e.m.f.

Φ is the total magnetic flux

t is the time taken to cut the flux

Motors

$n_s = \dfrac{f}{p}$ for synchronous speed (n_s)

where

f is the supply frequency

p is the number of pole pairs

$s = \dfrac{n_s - n_r}{n_s}$ for per unit slip where n_r is the rotor speed.

Colour	Significant figures	Multiplier	Tolerance
Silver	—	10^{-2}	± 10%
Gold	—	10^{-1}	± 5%
Black	0	1	—
Brown	1	10	± 1%
Red	2	10^2	± 2%
Orange	3	10^3	—
Yellow	4	10^4	—
Green	5	10^5	± 0.5%
Blue	6	10^6	± 0.25%
Violet	7	10^7	± 0.1%
Grey	8	10^8	—
White	9	10^9	—
None	—	—	± 20%

Resistor colour codes

Students studying the City & Guilds Course 185 and Course 200 have to be familiar with resistor colour codes. This form of identification is also used for capacitor markings and is found in BS 1852. It arose out of the difficulty of reading high resistive/ capacitive elements and also the difficulty of identifying them once mounted in position. Figure 1.0 shows a table of the resistor colour code used to overcome this problem. The colour bands represent signficant figures, multipliers and tolerance and the first thing to observe is that the bands are to one end of the resistor and should always be read from that end. The first band tells us the value between 1 and 9, e.g. red is 2 and the second band also tells us the value between 1 and 9, e.g. violet is 7. The third band tells us the number of noughts to be added to the digits 2 and 7 and it is very important for it to be read correctly otherwise a gross error could be made. Here the orange band is 3, thus making 27 000 ohms (or 27 kΩ). The fourth band is the tolerance or range within which the manufacturer guarantees the value of the resistor. In this case it is coloured gold and this tells us that its value can be 5% more or 5% less than the stated value.

Another method of determining resistor values is using a letter and digit code. In this method the code letters replace the decimal point. The letter R is used as the decimal point and the letters K,M,G and T are used as multipliers for 10^3, 10^6, 10^9 and 10^{12}. For

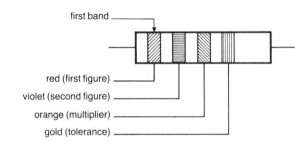

first band

red (first figure)

violet (second figure)

orange (multiplier)

gold (tolerance)

Fig. 1.0   Resistor colour codes.

example a resistor of 0.27 Ω is coded R27 and one which is of a 27.32 Ω is coded 27R23. A resistor of 100 Ω is coded 100R. In terms of other letters, a resistor of value 10 kΩ is coded 10K and one that has a value 10.33 kΩ is coded 10K33. A 1 M Ω resistor would be coded 1M0 and a 10 MΩ would be coded 10M. For tolerances, the letters B, C, D, F, G, J, K, M and N are used signifying ±0.1%, ±0.25%, ±0.5%, ±1%, ±2%, ±5%, ±10%, ±20% and ±30% respectively. If a 4.7 MΩ resistor has a tolerance of 20%, it would be coded 4M7M.

Transposition of formulae

Students will have noticed by now that there is a great number of formulae involved in electrical work and often to solve a particular problem it requires changing the subject of a formula for the required quantity. The first thing to appreciate in a formula is that the multiplication sign between term symbols is often omitted but it should always be returned when the terms are replaced with numbers.

Let us consider the Formula $Q = It$ which is explained in the next chapter. Each of the terms has a definite meaning and Q is found by multiplying together the product factors I and t. To make I the subject of the formula, t has to be removed. The only way this can happen is by dividing both sides of the formula by t.

Thus, if $\quad Q \;= It$
then $\qquad Q/t = It/t$

It will be seen that the t on the right-hand side can be cancelled out leaving Q/t on the left-hand side and I on the right-hand side.

Thus $\quad Q/t = I$
or $\qquad I \;= Q/t$

The same procedure can be repeated in order to find t. If there are more than two unknown product factors, just divide by those not required. Let us look at another formula, $R = \varrho l/a$. Say we were required to make a the subject of the formula. The simplest method here is to turn the formula upside down so that it becomes:

$1/R = a/\varrho l$

Now just multiply ϱl on the top line of both sides of the formula and again we will see that the ϱl on the right-hand of the equal sign will cancel and disappear leaving a on its own.

Thus $\quad \varrho l/R = \varrho la/\varrho l$
and $\quad \varrho l/R = a$
or $\qquad a = \varrho l/R$

It must be remembered that the transposing procedure requires *you* to make decisions but you must know the rules. Some of these are explained in the following examples.

Let us take a further example, such as $V = E - IR$. Say we had to find R. Here, we do not need to divide both sides by I, it is easier to make $-IR$ positive by taking it across the other side of the equal sign to the left-hand side.

Thus $\quad V + IR = E$

Now bring V across to the right-hand where it becomes $-V$

Hence $\quad IR = E - V$

Applying the procedure above, divide both sides of the equal sign by I since we want to make R the subject of the formula.

Thus $\qquad IR/I = (E - V)/I$
therefore $\qquad R = (E - V)/I$

One final example to show is the formula using $Z = \sqrt{(R^2 + X^2)}$. Here we are dealing with a square root sign and the first job is to re-arrange this formula in terms of Z^2

For example, if $\quad 5 \;= \sqrt{25}$
then: $\qquad\qquad 5^2 = 25$
Hence: $\qquad\qquad Z^2 = R^2 + X^2$

The square root vanishes. If X were required, then R^2 would become $-R^2$ on the left-hand side of the equal sign.

Thus $\qquad Z^2 - R^2 = X^2$

This can now be re-arranged to show X the subject of the formula and is found by square rooting the rest of the formula.

Therefore $\qquad X = \sqrt{(Z^2 - R^2)}$

The term R can also be found this way. It should be pointed out that both R and X will always be less than Z since the formula relates to Pythagoras's theorem and Z will always be the greater value:

Examples

Transpose the following formulae:

a) $A = \pi r^2$ $\qquad$ make r the subject of the formula
b) $X = 1/2\pi fC$ $\qquad$ make f the subject of the formula
c) $P = V^2/R$ $\qquad$ make V the subject of the formula
d) $R_1 = R_0(1 + \alpha t)$ make t the subject of the formula
e) $IR = E - Ir$ $\qquad$ make I the subject of the formula

Powers of a quantity (indices)

Algebra is frequently used to illustrate the rules for multiplying and dividing powers of a quantity. Some of these powers and their related quantities have

already been mentioned in transposition of formulae
but it will be seen from metric prefixes that quantities
or numbers may have positive or negative powers.
For example, the metric prefix 'M' meaning mega has
a positive index 6, i.e. 10^6. This should be read as ten
to the power of six and it indicates how many 10s
have to be multiplied together to make 1 000 000. In
the same way the metric prefix 'm' meaning milli has
a negative index -3 and this is written 10^{-3}. It is read
as ten to the power of minus 3. These negative
powers save you writing down the quantity as a
fraction, such as $1/10^3$ or $1/1000$ (one thousandth).
Students will have noticed that Appendix 8 of the
IEE Wiring Regulations incorporates time–current
characteristics of protective devices and that both
axes chosen for the graphs are labelled in powers of
ten. It is important, therefore, that the laws of
indices are fully understood. A range of these powers
of ten have already been given but also remember
that $10^1 = 10$ and $10^0 = 1$.

In multiplication, the index of the product of two
powers of the same quantity is the sum of their
separate indices. For example:

$$10^3 \times 10^6 = 10^{3+6} = 10^9$$
$$10^2 \times 10^{-3} = 10^{2-3} = 10^{-1}$$
$$10^6 \times 10^{-6} = 10^{6-6} = 10^0 = 1$$

In division, the index of the divisor must be
subtracted from the index of the dividend
(numerator). For example:

$$10^3 \div 10^6 = 10^{3-6} = 10^{-3}$$
$$10^2 \div 10^{-3} = 10^{2+3} = 10^5$$
$$10^6 \div 10^{-6} = 10^{6+6} = 10^{12}$$

In instances where metric prefixes are given, positive
indices are found abbreviating high values and
negative indices are found abbreviating very small
values. For example:

a) The Supergrid system opeates at 400 kV
   (i.e. 400×10^3 V)
b) The insulation resistance of a cable was
   measured to be 200 MΩ (i.e. $200 \times 10^6\ \Omega$)
c) The capacitor was marked 100 μF
   (i.e. 100×10^{-6} F)
d) The inductor was designed at 0.5 mH
   (i.e. 0.5×10^{-3} H).

Mensuration

This is concerned with the mathematical rules for
findings lengths, areas, volumes, etc. of figures. A
possible starting point might be to consider the
measurement of angles. With reference to Figure 1.1,
let us suppose the straight line OX starts from the
position OA and moves through the positions OB,
OC, OD and back to OA so that it describes a circle
about O in an anticlockwise direction. The line OX
generates the angle AOX as it passes round to
complete one revolution. In doing so the angle is
divided into 360 equal parts called degrees (360°).
One degree is divided into 60 minutes and one
minute is divided into 60 seconds. It will be seen that
the completed circle is divided into four equal parts
called quadrants and these create right-angles, with
each one containing 90°.

If you place a protractor over the angle AOX in
the first quadrant you will find this to be 30°. Some
further terms relating to the geometry of the circle
are: OA is called the radius; CA is called the
diameter; AX is called the arc created by the angle
AOX and for the completed circle it is called the
circumference; AB is called a chord as it divides the
circle into two parts called segments (the one in the
first quadrant is called the minor segment). Finally,
the part which is bounded by two radii and an arc
between them is called a sector (the quadrants are
sectors). Within the first quadrant is created a right-
angle triangle (AOB) and by using your protractor
over the other angles you will notice that they both
contain 45°. This tells us that the number of degrees
in any triangle will not exceed 180°.

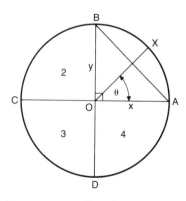

Fig. 1.1   Measurement of angles.

The right-angled triangle in Figure 1.1 represents a square cut in half and the diagonal line AB is the longest line called the hypotenuse. Whilst there are numerous types of triangle to be found, in alternating current theory we use the trigonometrical ratios of the right angle to explain the concept underlying such terms as impedance and power factor. This is mainly because a.c. is generated in conductors which are part of a rotating system that can easily be explained by several well-known mathematical principles. Before the rules relating to trigonometry are mentioned, let us first deal with the simple theorem of Pythagoras. This theorem can be stated as follows:

> 'in a right-angled triangle the square on the hypotenuse is equal to the sum of the squares on the other two sides.'

Draw a right-angled triangle with sides measuring 3 cm, 4 cm and 5 cm and complete the squares as indicated by the theorem (see Figure 1.2). It will be seen that $AB^2 = AO^2 + BO^2$

$$\text{therefore} \quad 5^2 = 3^2 + 4^2$$
$$\text{and} \quad 25 = 9 + 16$$

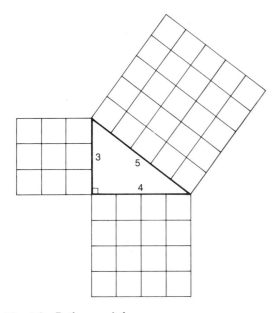

Fig. 1.2 Pythagoras' theorem.

Examples

1 A ladder 4 m long rests against a wall of a house so that the foot of the ladder is 1 m from the foot of the wall. How far up does it reach?

Solution

Providing the wall is at right-angles to the ground we can use the theorem described above. Thus, the ladder length must represent the hypotenuse (say AB) and the base length (say AO). The question asks us to find the perpendicular length (say BO).

$$\text{If} \quad AB^2 = AO^2 + BO^2$$
$$\text{then} \quad BO^2 = AB^2 - AO^2$$
$$\text{Thus} \quad BO^2 = 4^2 - 1^2$$
$$= 15$$
$$\text{and} \quad BO = \sqrt{15} = \underline{3.87 \text{ m}}$$

2 A resistor of 15 ohm is connected in series with a inductor of reactance 20 ohm. Ignoring any resistance in the inductor, what is the impedance of the circuit?

Solution

The principles behind this question have not yet been explained but the procedure in solving it requires no transposition of formula.
Let impedance (Z) be the hypotenuse (say AB) and the resistor (R) be the side (say AO). The inductor (X) can be represented by the other side (say BO).

$$\text{Thus} \quad Z^2 = R^2 + X^2$$
$$= 15^2 + 20^2$$
$$= 225 + 4000$$
$$= 625$$
$$\text{and} \quad Z = \sqrt{625}$$
$$= \underline{25 \ \Omega}$$

A brief mention will now be made to some other types of triangle as well as the plane figures called parallelograms and quadrilaterals. The triangle is of course a figure enclosed by three straight lines. An *acute-angled triangle* is one having three acute angles, i.e. each of them is less than a right angle; an *obtuse-angled triangle* has one of its angles greater than a right angle; an *isosceles triangle* has two of its sides equal in length as well as having two angles equal; and an *equilateral triangle* is a triangle having three sides and three angles equal. Let us consider the equilateral triangle in more detail. This triangle is shown in Figure 1.3 and students should construct it themselves making each side 6 cm long – for accuracy use compasses and check that all the angles are 60°. Now letter the figure as shown. The line marked h is called the perpendicular since it is at right angles to the base line CB. The length of this line can be measured using a ruler but since the theorem of Pythagoras has been explained, you will find it more accurate to calculate it, being approximately 5.2 cm long. Now the area of any triangle can be found by multiplying half the base measurement (CB) by the perpendicular height (h). Thus:

$$\text{Area } (A) = \tfrac{1}{2}\text{ base} \times \text{height}$$
$$= \tfrac{1}{2}\,(6 \times 5.196)$$
$$= \underline{15.59 \text{ cm}^2}$$

An alternative method for determining the area of any triangle is using the formula:

$$A = \sqrt{[s\,(s-a)\,(s-b)\,(s-c)]}$$
$$\text{where} \quad s = \tfrac{1}{2}\,(a + b + c)$$

For example, in the above question,
$$s = \tfrac{1}{2}\,(6 + 6 + 6) = 9$$
$$\text{therefore} \quad A = \sqrt{[9\,(9-6)\,(9-6)\,(9-6)]}$$
$$= \underline{15.59 \text{ cm}^2}$$

Other plane figures include types of *quadrilateral*. These are figures bounded by four straight lines. Those in which the opposite sides are parallel and equal to each other are called *parallelograms*. Parallelograms in which all the angles are at right angles are called *rectangles*. A *square* is a rectangle in which the sides are of equal length. A parallelogram with equal sides but the angles are not at right angles is called a *rhombus*. One further plane figure to mention is the *trapezium* and this is a quadrilateral with one pair of opposite sides parallel. *Polygons* (figures bounded by any number of straight lines) will not be discussed.

The area of squares and rectangles can easily be found by multiplying length × breadth but in a parallelogram the figure needs to be transformed into a rectangle since measurement is required of its perpendicular height. Figure 1.4 shows a typical parallelogram labelled ABCD. With perpendiculars

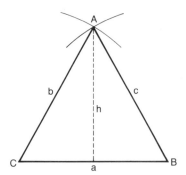

Fig. 1.3 Equilateral triangle.

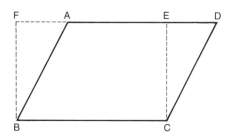

Fig. 1.4 Parallelogram.

drawn BF and CE, the figure can be pushed back to become a rectangle and its area ascertained by multiplying its base by the perpendicular height.

Thus   Area = AD × CE

Note: If a diagonal line were drawn between A and C, the parallelogram could be treated as two separate triangles and the area found by

Area = 2 × ½AD × CE.

When finding the area of trapeziums, try to transform them into rectangles or even triangles and add the separate areas together. The circle shown in Figure 1.1 is by definition a plane figure contained by a line described as the circumference. The ratio of the circumference (C) to the diameter (d) is denoted by pi (π). Pi is the circle's constant and is quite commonly taken as 3.142. The circumference of any circle can be found from the formula: $C = \pi d$. Since the diameter is twice the radius,

then $C = 2\pi r$

In terms of area, students are not expected to know the mathematical proof of the circle but it will be sufficient to say that the length of an arc may be measured approximately by dividing it into very short portions, and by measuring these lengths as if they were straight. The area of the circle can be found by adding together all the areas of the small isosceles triangles thus formed (i.e. the sum of all the ½ base × height). Hence the area of a circle can be found by the formula:

$$\text{Area} = \tfrac{1}{2} \times 2\pi r \times r = \pi r^2$$

Example

A copper busbar measures 5 cm in diameter. What is the length of its circumference and cross-sectional area?

Solution

From the above formula
$$\begin{aligned} C &= \pi d \\ &= 3.142 \times 5 \\ &= \underline{15.71 \text{ cm}} \end{aligned}$$

Also
$$\begin{aligned} A &= \pi r^2 \\ &= 3.142 \times 2.5^2 \\ &= \underline{19.63 \text{ cm}^2} \end{aligned}$$

Note: It should be pointed out that the area of a circle (πr^2) can be expressed in terms of its diameter (d) instead of its radius (r).

Since $r = d/2$
and $r^2 = (d/2)^2$
  $= d^2/4$
then $A = \pi d^2/4$

Using this formula is often much quicker than getting involved with radius measurements as will be seen in the following example.

Example

A metal conduit has an outside diameter (D) of 20 mm and an inside diameter (d) of 15 mm. What is the rim area of the conduit?

Solution

This example is solved by subtracting the conduit's inside area from its outside area. The formula used is

that for finding the area of an annulus or circular ring. Thus:

$$\begin{aligned} A &= (\pi D^2/4) - (\pi d^2/4) \\ &= (\pi/4)(D^2 - d^2) \\ &= 0.785(20^2 - 15^2) \\ &= \underline{137.4 \text{ mm}^2} \end{aligned}$$

Students should also know how to find the volume of regular solids such as the square prism and cylinder. The former is a solid with two plane ends of the same size and shape whereas the latter is a solid having two parallel circular ends (see Figure 1.5). In both cases illustrated, the volume (V) is the product of the base area (A) and height (h), i.e.

$$V = Ah$$

If the volume is required of a hollow cylinder the following formula is used:

$$V = (\pi h/4)(D^2 - d^2)$$

If the total surface area (S) of a cylinder is required the formula is

$$S = 2\pi r (h + r)$$

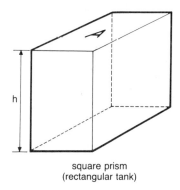

square prism
(rectangular tank)

circular prism
(cylinder)

Fig. 1.5   Regular solids.

Example

A solid copper busbar is 0.5 m long and 2 cm in diameter. What is its total surface area and volume?

Solution

$$S = 2\pi(50 + 1) = \underline{320.48 \text{ cm}^2}$$
$$V = \pi 50 = \underline{157 \text{ cm}^3}$$

Irregular figures

It is not normally required of students to find the areas of irregular figures but they are expected to know how to determine the average value and root mean square value of an a.c. sinusoidal wave. Figure 1.10 shows how an a.c. sinewave is produced. It will be seen that its average value over the complete cycle is zero but for half a cycle, as would be obtained from a rectified supply, it is found to be the area under the curve divided by the length of base. One method of obtaining this value is using the *mid-ordinate rule* where the lengths of mid-ordinates are measured from the base axis to their points of intersection on the curve. By dividing the number of mid-ordinates (n), the average value is found. Theoretically this should be 0.637 × maximum value. For an a.c. voltage, the formula to use is:

$$\text{average value} = (e_1 + e_2 + \ldots e_n)/n$$

This value in a.c. theory is of little importance. It is the effective value or root mean square value that is important and which is indicated by measuring instruments. It is expressed in terms of a direct current which produces the same heating effect in a resistor. It is found by the formula:

$$\text{r.m.s. value} = \sqrt{(e_1^2 + e_2^2 + \ldots e_n^2)/n}$$

This value should theoretically be 0.707 × maximum value.

Example

The maximum value of an a.c. sinewave voltage is 339.46 V and that of an a.c. sinewave current 20 A. Determine their r.m.s. values?

Solution

$$V_{\text{r.m.s.}} = 0.707 \times 339.46 = 240 \text{ V}$$
$$\text{and} \quad I_{\text{r.m.s.}} = 0.707 \times 20 \quad = 14.17 \text{ A}$$

It will be noticed that the sinewave in Figure 1.10 covers only one cycle. In practice, the public supply is providing 50 cycles every second, called 50 Hertz. One cycle per second is called the periodic time (T) and equals $1/f$. For 50 Hz, one cycle takes 0.02 s.

Trigonometry

Part II students will almost certainly come across trigonometrical ratios in their studies and in the particular case of the right-angled triangle, trigonometry proves a very useful method of finding unknown angles and sides. The three sides of the triangle are labelled hypotenuse, opposite and adjacent and the two unknown angles are often given the Greek letters theta (θ) and phi (ϕ). *It is very important to remember that the adjacent side for the angle ϕ is the opposite side for the angle θ. In the same way the adjacent side for the angle θ is the opposite for the angle ϕ.* The ratios are expressed in terms of sine, cosine and tangent and they are abbreviated sin, cos and tan respectively.

For the angle ϕ the ratios are:

> $\sin \phi$ = opposite/hypotenuse
> $\cos \phi$ = adjacent/hypotenuse
> $\tan \phi$ = opposite/adjacent

For the angle θ the ratios are:

> $\sin \theta$ = opposite/hypotenuse
> $\cos \theta$ = adjacent/hypotenuse
> $\tan \theta$ = opposite/adjacent

Values of sin, cos and tan can be found in trigonometrical tables or they can be obtained from any scientific calculator. Since we are dealing with a right-angled triangle, these values (expressed in degrees) will range between 0° and 90°.

Let us now consider the right-angled triangle mentioned earlier which has sides 3 cm, 4 cm and 5 cm. This is shown again in Figure 1.6. The opposite side to the angle ϕ is 3 cm while the adjacent side to the angle ϕ is 4 cm and obviously the hypotenuse is 5 cm. We know that both angles ϕ and θ when added together equal 90° and one way of finding out their values is by using a protractor. Trigonometry allows us to be a little more accurate than this and also there is no need to draw scaled diagrams.

Let us now make ratios of the angle ϕ, e.g.

> $\sin \phi$ = 3/5 = 0.6
> $\cos \phi$ = 4/5 = 0.8
> $\tan \phi$ = 3/4 = 0.75

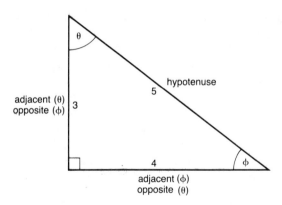

Fig. 1.6   Trigonometry ratios.

If reference is now made to trigonometrical tables or a calculator you will find that:

$$\sin^{-1} 0.6 = 36.87°$$
$$\cos^{-1} 0.8 = 36.87°$$
$$\text{and } \tan^{-1} 0.75 = 36.87°$$

Note: The use of the terms $\sin^{-1}$, $\cos^{-1}$ and $\tan^{-1}$ is simply to denote the angles being referred to them.

It will be seen that each ratio has given the answer 36.87° which it obviously must, and it should be noted that the numbers after the decimal point represent part of a degree (87/100) and not minutes. From this result it is not too difficult to find that the angle θ is 53.13° Let us now check this by using the ratios relating to angle θ, e.g.

$$\sin θ = 4/5 = 0.8 \quad \text{and } θ = 53.13°$$
$$\cos θ = 3/5 = 0.6 \quad \text{and } θ = 53.13°$$
$$\tan θ = 4/3 = 1.333 \text{ and } θ = 53.13°$$

The above ratios are the basic tools required to solve any right-angled triangle.

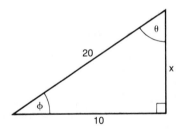

Fig. 1.7   Right-angled triangle.

Example 1

Figure 1.7 shows a right-angled triangle with hypotenuse side 20 units and adjacent side to the angle φ being 10 units. Find the two unknown angles and the unknown side.

Solution

$$\cos φ = \text{adjacent/hypotenuse}$$
$$= 10/20 = 0.5$$
$$φ = 60°$$
therefore $\quad θ = 30°$
Since $\quad \sin φ = \text{opposite/hypotenuse}$
then opposite side $\quad = \text{sine } φ × \text{hypotenuse}$
$$= \sin 60° × 20$$
$$= 0.866 × 20$$
$$= \underline{17.32 \text{ units}}$$

Let us check the answers

$$\sin θ = 10/20 = 0.5$$
$$θ = 30°$$
$$\tan θ = \text{opposite/adjacent}$$
then adjacent side $\quad = \text{opposite/tan } θ$
$$= 10/\tan 30°$$
$$= 10/0.577$$
$$= \underline{17.32 \text{ units}}$$

Example 2

Figure 1.8 shows a right-angled triangle having sides marked R, X and Z. If the angle φ = 45° and side $X = 100 \, Ω$, what are the ohmic values of R and Z?

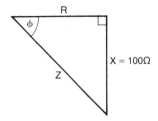

Fig. 1.8   Right-angled triangle.

Solution

The triangle represents ohmic values associated with resistance (R), reactance (X) and impedance (Z) and the angle φ when related to the cosine indicates the power factor of the circuit. To solve the problem using trigonometry then:

$$\sin \phi = \text{opposite/hypotenuse}$$
$$= X/Z$$
therefore
$$Z = X/\sin \phi$$
$$= 100/0.707$$
$$= \underline{141.4 \ \Omega}$$

Since
$$\tan \phi = \text{opposite/adjacent}$$
$$= X/R$$
then
$$R = X/\tan \phi$$
$$= 100/1$$
$$= \underline{100 \ \Omega}$$

Note: It will be seen that $R = X$ in value. This is because the angle 45° tells us that the right-angled triangle is half a square. Of interest is the diagonal (Z) which in a square figure is always 1.414. ($\sqrt{2}$) longer than its side measurement. This means that the value of Z when checked is found to be 1.414 × 100 = 141.4 Ω.

Example 3

For safety reasons before being climbed, a ladder resting against a perpendicular wall should obey the 4:1 rule (i.e. 1 m out for every 4 m in height). What are the angles created by this rule?

Solution

Let the ladder base angle be called ϕ and the other θ.

Thus
$$\tan \phi = \text{opposite/adjacent}$$
$$= 4/1 = 4$$
and
$$\phi = \underline{75.96°}$$
to make 90°
$$\theta = \underline{14.04°}$$

Check angle θ, i.e. $\tan \theta = 1/4 = 0.25$
$$\theta = 14.04°$$

Graphs and Charts

A graph is a diagrammatic way of conveying data with the purpose of comparing related quantities. It is generally plotted between axes at right angles to each other. The base (x-axis) is called the *abscissa* and the perpendicular (y-axis) is called the *ordinate*. It is important to choose suitable scales for both axes and these should be clearly marked and given titles of the quantities they represent. The scales need not start from zero but they should be extended sufficiently so that the graph is not confined to a small part of the graph paper.

If a graph were drawn in the first quadrant of Figure 1.1, both the x and y axes would be given positive values. In the second quadrant the x-axis has a negative value while the y-axis remains positive. If a graph was portrayed in the third quadrant both axes would have negative values but in the fourth quadrant the x-axis returns to a positive value while the y-axis remains negative.

Let us now consider some of the common types of graph found in electrical science studies. Probably one of the first graphs drawn by students in the first year will be the two axes graph of Ohm's Law whereby a linear resistor is used to demonstrate the relationship between potential difference and current. The graph of these two quantities should show a direct variation and be a sloping straight line rising from zero as shown in Figure 1.9. It tells us that the potential difference across the ends of the resistor is proportional to the steady current passing through it. Graphs which start off high and then slope downwards denote inversely proportional characteristics such as found in the relationship between resistance and cross-sectional area and between capacitive reactance and frequency.

Graphs which result in straight lines, either horizontally or vertically, indicate that one of the quantities is unaffected by any variation in the other

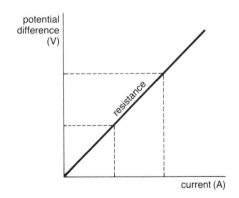

Fig. 1.9 Graph showing Ohm's Law.

quantity. Some graphs rise up proportionally and then start to bend over and gradually level out. These often show that a point of saturation has been reached where the *y*-axis no longer increases at the same rate as the *x*-axis. An example of this is found in *B–H* magnetization curves.

Since reference has already been made to the circle, the graph of a sinewave over one complete cycle of 360° is shown in Figure 1.10 and it illustrates how an induced e.m.f. is created by a rotating conductor situated in a uniform magnetic field. The *x*-axis refers to the periodic time of this wave and the scale will be marked out by twelve intervals, each being 1 cm apart and representing 30°. The *y*-axis refers to the induced e.m.f. and is given an overall scale measurement of 10 cm. It will be seen that this axis is really 5 cm in the positive direction and 5 cm in the negative direction – controlled by the circle's radius. The sinewave produced is cyclic in nature, reaching two maximum values, one at 90° and the other at 270°. The thirteen points marked on the graph are called *instantaneous points* and it will be noticed that the induced e.m.f. does not exist at 0°, 180° and 360°. Other properties of this a.c. sinewave will be discussed later in other chapters.

Students are not required to produce circle charts (pie charts) or Venn diagrams but they need to know the procedure for drawing bar charts which are similar to histograms used in statistics in which frequency distributions are illustrated by rectangles. Bar charts have a wide use in the construction industry to show changes occurring in projects, such as between work operations and the length of time it takes for their completion. These charts are mentioned in Volume 1, *Theory and Regulations*. They are drawn from left to right across graph paper with the parameter representing a time scale in days or weeks. Work activities are listed below each other and their inter-relatedness compared. Some of these activities cannot start before others and unlike normal histograms they do not all start on the same line.

Ratio, proportion and percentage

One final topic in this chapter concerns the application of problems associated with these three methods. It is common to speak of a ratio as the numerical relation one quantity bears to another of

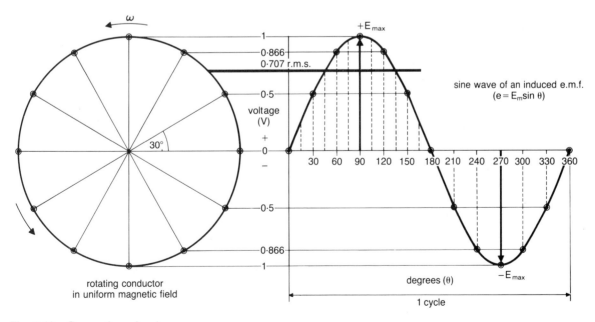

Fig. 1.10 Generation of a sine wave.

the same kind of units. The sides of a right-angled triangle have already been mentioned in terms of trigonometrical ratios and they are simply obtained by dividing one quantity by the other. For example, 100 V divided by 60 V will give a ratio of either 100:60, 50:30, 10:6 or even 5:3. In practice, ratios should always be written in the easiest form and it should be pointed out that 5:3 is not the same as 3:5. A transformer may have a step-down voltage ratio of 5:3 but if it was stated 3:5, it would be a transformer having a step-up voltage ratio.

Where there is equality between two ratios it is called proportion ($\propto$). We have already seen that the circumference of a circle is proportional to its diameter. When two circles are compared, we can say that the ratio of circumferences of the first circle and second circle equals the ratio of diameters of the first circle and second circle, i.e. $C_1/C_2 = d_1/d_2$.

It will be seen in Chapter 2 when dealing with resistance connections, that current is shared in different parts of a parallel circuit. These branch currents will be a proportional amount of the supply current, satisfying Kirchhoff's first law which states:

Current flowing towards a junction is equal to the current flowing away from that junction.

The usual way of writing down quantities which are proportional to each is to insert the symbol between them. For example, the resistance (R) of a wire is proportional to its length (l) and inversely proportional to its cross-sectional area (A). This is normally written:

$R \propto l/A$

A percentage is the fractional part of a quantity with 100 as the denominator. For example, $3/4 = 0.75 = 75/100 = 75\%$. In cable problems we are often asked to find the voltage drop (V) in circuits. The maximum allowance is based on 2.5% of the declared voltage. If the supply voltage were 240 V, then 2.5% of 240 V is written:

$V = 2.5/1000 \times 240 = 6$ V

Often manufacturers of machines and switchgear express the operation of their equipment in values above 100%, e.g. 150%, 200%, 300% etc. This approach is often used to provide information about the apparatus overcurrent capability and since

100% = 1, the above should be interpreted as 1.5, 2 and 3 times the normal conditions. As another example, a water heater which is 90% efficient must have 10% losses. It is becoming common practice to interpret results in per unit values rather than in percentage form. For example, a motor's rotor might have a percentage slip of 5% which is written as 0.05 per unit.

Equations

A statement of equality between known and unknown quantities is called an equation. For example, $5y = 10$ is an equation and is only true when $y = 2$. The symbol y is chosen randomly and like many other letters of the alphabet it is used to represent a quantity. For example, l is used to represent length, A is used for area, V is used for volume and you have already seen the many symbols used to represent electrical terms.

Consider a 3 m length of conduit cut into three pieces. The second piece is twice as long as the first while the third piece is 40 cm shorter than the first. If 20 cm is allowed for wastage, find the length of the three pieces. The solution to this problem is found by creating an algebraic equation from the information, For example:

Let the first piece be called x
Let the second piece be called $2x$
Let the third piece be called $x-40$

Since 20 cm is wastage, the conduit used is 280 cm. Hence the following equation can be written:

$$x + 2x + (x - 40) = 280$$
$$4x - 40 = 280$$
$$4x = 320$$

Thus, first piece is $\quad x = \underline{80\ cm}$
and second piece is $\quad 2x = \underline{160\ cm}$
and third piece is $\quad 80 - 40 = \underline{40\ cm}$

It frequently happens that the solution to a problem involves finding two unknowns and one method is to use a simultaneous equation which is using two equations with similar data and connecting them together by addition or subtraction to solve. Consider:

$$6a + 10b = 2 \qquad (1)$$
$$9a + 6b = 12 \qquad (2)$$

In equation (2) multiply by 4 and in equation (1) multiply by 6

$$36a + 24b = 48 \quad (3)$$
$$\text{and} \quad 36a + 60b = 12 \quad (4)$$

now subtract eqn (4) from eqn (3), i.e. change signs of (4) and add.

$$\text{thus} \quad -36b = 36$$
$$\text{therefore} \quad b = -1$$

substitute b in eqn 1,

$$\text{i.e.} \quad 6a + 10(-1) = 2$$
$$6a = 12$$
$$a = 2$$

The method just described is called the *elimination method* which leaves the equation containing only one unknown, the value of which can be found in the usual manner. Another method is called the *substitution method* and is shown below:

Consider the previous example where:

$$6a + 10b = 2 \quad (1)$$
$$\text{and} \quad 9a + 6b = 12 \quad (2)$$

In equation (1) make a the subject of the formula and substitute its value in equation (2).

Hence $a = (2 - 10b)/6$
and in eqn. (2) $9(2 - 10b)/6 + 6b = 12$
$$3 - 15b + 6b = 12$$
$$-9b = 12 - 3$$
$$b = -1$$

The value of a can be found by substituting b back into equation (1) as previously dealt with.

In Chapter 2 it will be seen how Kirchhoff's laws help to explain the distribution of current and voltage around a closed loop circuit. Simultaneous equations are used to solve this type of problem. One further example will now be given to illustrate this:

The p.d.s in a circuit when added together equal the sum of the e.m.f.s and may be expressed as: $\Sigma E = \Sigma IR$. In a closed circuit, two equations are derived based on this fact.

$$\text{Let} \quad E_1 = I_1R + I_2R \quad (1)$$
$$\text{and} \quad E_2 = I_1R + I_2R \quad (2)$$

If values are given to the two equations, then the two unknown currents can be solved.

Thus $33 = 6I_1 + 3I_2 \quad (1)$
and $19 = 13I_1 - 4I_2 \quad (2)$
By multiplying (1) by 4 $132 = 24I_1 + 12I_2 \quad (3)$
and multiplying (2) by 3 $57 = 39I_1 - 12I_2 \quad (4)$
By addition of both eqs. $189 = 63I_1$
therefore $I_1 = 3$ A

By substituting this value in equation (1), then $I_2 = 5$ A

Exercise 1

1   Convert the following:
  a) 0.35 ohms into milliohms
  b) 750 millijoules into joules
  c) 255 kilowatts into megawatts
  d) 400 000 volts into kilovolts
  e) 0.022 millifarads into microfarads

2   Write down the Greek symbols for the following terms and briefly explain their use.
  a) efficiency
  b) pi
  c) resistivity
  d) proportional to
  e) equal to or greater than

3   a) Transpose the formula $A = \pi d^2/4$ to find d
  b) Calculate the diameter of a 2.5 mm^2 cable.

4   Solve a) $10^6 \times 10^3 \times 10^2$
  b) $10^{-5} \times 10^0 \times 10^3$

5   a) Using Pythagoras's theorem, determine the value of V_s in Figure 1.11.
  b) Using trigonometry, determine the angle ϕ

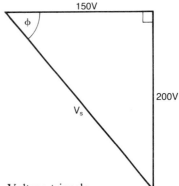

Fig. 1.11   Voltage triangle.

6   *a*) Determine the area of a triangle having
        sides 3 cm, 4 cm and 5 cm
     b) What type of triangle do the
        measurements suggest?
7   What is the volume of a copper busbar 3 m in
     length having an inside diameter 2.4 cm and
     outside diameter 3 cm?
8   The result table below shows the voltage and
     current readings obtained in an experiment to
     find the value of an unknown resistor. Plot the
     results on graph paper and determine the
     resistor's value from the slope of the line.

V	4.4	10	14	18.8	22
I	2.0	4	6	8	9

9   Draw the sinewave shown in Figure 1.10 over
     half a cycle and from it determine its average
     and root mean square values using the mid-
     ordinate rule.
10   A 20:1 step-down transformer has a primary
     voltage of 6.6 kV. What is its secondary
     voltage?

CHAPTER TWO

Basic circuit theory

After reading this chapter you will be able to:

1 State a number of electrical formulae associated with basic circuit theory.

2 Know various electrical laws relating to resistance and inductance.

3 Perform calculations involving resistance, inductance and capacitance.

4 State the difference between energy and power, and perform energy and power calculations.

5 State the magnetic effects of current and apply various rules to indicate current and magnetic force direction.

6 State the meaning of electromagnetic induction and apply the principles to a.c. and d.c. generation.

7 Draw diagrams which illustrate various rules and laws associated with basic circuit theory.

8 State the chemical effects of current and apply the laws relating to electrolysis.

9 Know the basic theory and construction of secondary cells and carry out calculations on cell connections.

10 Know the procedure for charging secondary cells through a rectifier circuit.

Electricity

Electricity is concerned with the energy present in protons and electrons. It is these and other small particles that constitute the essential ingredients of all atoms which make up matter in the form of solids, liquids and gases.

An atom is not a solid entity, it is often likened to a small solar system having a central nucleus containing positively charged particles called *protons*. Around the nucleus orbit negatively charged particles called *electrons* and these are generally attracted to the nucleus, keeping the atom in a neutral state.

Elements of different substances are given atomic numbers based on how many protons they possess. For example, hydrogen has an atomic number of 1 since it has only 1 proton and 1 electron. Oxygen has an atomic number of 8 whereas for copper it is 29 (see Figure 2.1) and for uranium it is 92.

At times an atom may lose or gain an electron and when this happens it is called an *ion* since it is electrically charged. If it loses an electron it is called a *positive ion* whereas if it gains an electron it is called a *negative ion*. The process of forming ions is called *ionization* and is a term often used in the operation of gas discharge lamps.

For a material to conduct electricity it must allow charges to flow through it. It is not unusual to find positive and negative charges flowing in opposite directions, e.g. as ions in gases or liquid conductors.

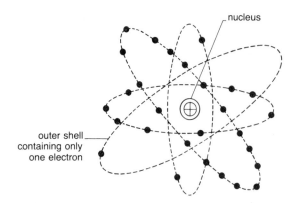

Fig. 2.1 Copper atom showing orbiting negatively charged electrons.

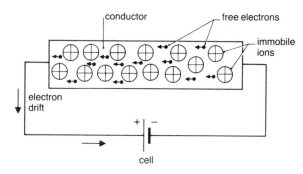

Fig. 2.2 Movement of 'Free electrons'.

In solid electrical conductors, however, it is only electrons which are able to flow and because of their negative charges they need to be attracted towards a potential source such as the positive pole of a battery. Figure 2.2 helps to explain this theory*.

Unfortunately, the charge of an electron is too small for practical measurement, so a larger unit called a *coulomb* is often used. One coulomb is equivalent to 6.3×10^{18} electrons. This is the quantity of electricity (Q) crossing a section of the conductor in a time (t) of 1 s and it defines the term *electric current* (I). Expressed as a formula this can be written as:

$$I = Q/t \qquad (2.1)$$

where *I* is the current in amperes (A)
  Q is the charge in coulombs (C)
  t is the time in seconds (s)

The simplest analogy of an electric circuit is to consider a garden water pipe connected to a tap. The rate of water flowing out of the end of the pipe will depend upon the initial tap pressure and the flow of water will be restricted by the inner walls of the pipe, particularly where bends and kinks occur. If there are many restrictions this will be noticeable by a pressure drop as the water comes out of the end of the pipe. In the same way, current flows through conductors by means of an electric pressure from a battery or generating source. This source of electric pressure is

called *electromotive force* (e.m.f.) and provides the energy to move current through a circuit. The electromotive force is referred to as the *supply voltage* and for a stable supply the current allowed to flow is determined by resistance in the circuit conductors. There will be a pressure drop across different parts of the circuit and this is referred to as *potential difference* (p.d.).

Unlike the water pipe analogy, the electric circuit needs a 'go' and 'return' conductor and these must have low resistance values. Most solid conductors satisfy this requirement with copper and aluminium, the economic favourites for this purpose. Since there is a danger of bare conductors discharging current to earth, the practice is to either isolate them or insulate them. Good insulators are those materials having no 'free electrons' available and within normal temperature limits they cannot conduct. Rubber, pvc, ebonite, porcelain and glass are all typical insulators having extremely high resistance values.

It should be pointed out that some materials are neither good conductors nor good insulators, their electrical properties are somewhere between both low and high values of resistance. Materials like this are called *semiconductors* such as germanium and silicon and find considerable use as rectifier elements.

Ohm's Law

This is a law named after Georg Ohm (1787–1854). It states that:

The ratio of potential difference (*V*) between the ends of a conductor and the current (*I*) flowing in the conductor is constant. This ratio is termed the resistance of the conductor.

*Conventional current flow from a positive supply terminal is often used instead of electron flow and students must not be confused by this.

The temperature of a conductor must not change if the law is to be obeyed. It can then be stated as: The current flowing in a circuit is directly proportional to the potential difference (V) and inversely proportional to the circuit resistance (R). Expressed as a formula this statement becomes:

$$I = V/R \qquad (2.2)$$

where I is the current in amperes (A)
V is the potential difference (V)
R is the circuit resistance (Ω)

If the above conditions are satisfied, then increasing the p.d. will increase the current in the same proportion as was explained in the previous chapter. The formula (2.2) is often used for finding current values such as in the selection of a suitable fuse size or the selection of a suitable conductor size. When re-arranged, the working resistance of a component or the circuit voltage drop can be found.

Resistance factors

Experiments have shown that for a particular conductor material in a constant temperature its resistance is:

(a) directly proportional to its length (see Figure 2.3(a)) and

(b) inversely proportional to its cross-sectional area (see Figure 2.3(b)).

Both length and cross-sectional area are referred to as the conductor's dimensions. Another factor which has to be considered is *resistivity* since some materials conduct electricity more easily than others. It is a constant for a particular material and is defined as the resistance between the opposite faces of a unit cube of the material. Some common resistivity values are as follows, based on a temperature of 20°C.

Annealed copper	0.0172 $\mu\Omega$m
Hard-drawn aluminium	0.0285 $\mu\Omega$m
Rolled brass	0.09 $\mu\Omega$m
Tungsten	0.056 $\mu\Omega$m

The above resistance dimensions can be expressed as:

$$R = \varrho l/A \qquad (2.3)$$

where R is the resistance (Ω)
ϱ is the resistivity (Ωm)
l is the length (m)
A is the cross-sectional area (m^2)

The factor which considers the effects of temperature is called the *temperature coefficient of resistance* (α) and it relates to the fractional increase per degree Celsius of resistance at 0°C. This temperature coefficient can be expressed as:

$$\alpha = \frac{R_1 - R_0}{R_0 t_1} \qquad (2.4)$$

where α is the temperature coefficient of resistance ($\Omega/\Omega/°C$)
R_1 is the resistance at temperature t_1
R_0 is the resistance at 0°C
t_1 is the temperature (°C)

Most metal conductors increase in resistance when the temperature increases and they are said to have *positive* temperature coefficients of resistance, whereas most insulators have *negative* temperature coefficients of resistance. Included among these are electrolytes, liquids and also carbon. Figure 2.4 shows the variation in resistance of a good conductor material such as copper. If the graph of this metal extended backwards to meet the base line its temperature would be −235°C. It is worth noting that it has a melting point of 1084°C and it is not too difficult to appreciate that its resistance value will increase considerably over this range of temperature. Some typical values of α per °C at 0°C are:

Copper	0.004 28 $\Omega/\Omega/°C$
Aluminium	0.003 9 $\Omega/\Omega/°C$
Carbon	−0.000 5 $\Omega/\Omega/°C$

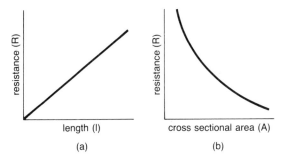

(a)   (b)

Fig. 2.3 Resistance dimensions.

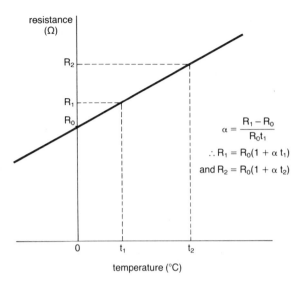

$$\alpha = \frac{R_1 - R_0}{R_0 t_1}$$

$$\therefore R_1 = R_0(1 + \alpha t_1)$$

and $R_2 = R_0(1 + \alpha t_2)$

Fig. 2.4 Variations of resistance and temperature.

Resistor connections

One method of finding the ohmic value of a resistor is to connect an ammeter and voltmeter in its circuit. The ammeter will directly measure the current flowing and should always be placed in series with the component. The voltmeter directly measures potential difference and should always be connected across components or across live terminals. The working value of resistance is found by dividing the voltmeter reading by the ammeter reading.

If a number of lamps are connected end-to-end in the form of a chain, they are said to be connected *in series*. In this mode of connection, the same current leaving the supply source passes through each lamp and the sum of all the p.d.s across them will equal the supply voltage. Christmas tree 'fairy lights' are connected like this and although the supply will be at 240 V a.c., each lamp requires a voltage of only 12 V. It is important to realize that if an open circuit occurred in the chain, then 240 V would appear at that point.

If, however, the 'fairy lights' were re-connected such that each side of a particular lamp shares a common connection with one side of all the other lamps, then they are said to be connected *in parallel*. In this mode of connection, each lamp receives the full supply voltage across it and it therefore needs to

be rated the same as the supply voltage. A further method of connection is called *series–parallel* which is basically a combination arrangement of the two just described.

Before discussing these methods in detail, it should be pointed out that Ohm's law applies reasonably well to most parts of the electric circuit but besides this law there are two further laws associated with current distribution and voltage distribution. These are called Kirchhoff's laws after Gustav Kirchhoff (1824–87). The first law can be interpreted as:

> The total current flowing towards a junction is equal to the current flowing away from that junction

and the second law interpreted as:

> in a closed circuit, the p.d.s of each part of the circuit are equal to the resultant e.m.f. in the circuit.

The series circuit is a good example of Kirchhoff's second law while the parallel circuit is a good example of the first law. Both arrangements will now be discussed.

Resistors in series

Figure 2.5 shows a typical series where R_1, R_2 and R_3 are three resistors and V_1, V_2, and V_3 are three voltmeters connected across each component measuring the respective p.d.s. The ammeter (A) indicates the current taken from the supply by the three resistors. Applying Ohm's law to all parts of the circuit, then:

$$V_1 = IR_1, \; V_2 = IR_2, \; V_3 = IR_3, \text{ and } E = IR_e$$

where E is the e.m.f. or supply voltage and R_e is the equivalent circuit resistance

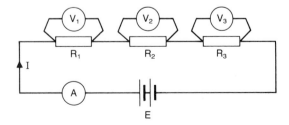

Fig. 2.5 Series circuit (common current).

Applying Kirchhoff's second law:

$$E = V_1 + V_2 + V_3$$
$$IR_e = IR_1 + IR_2 + IR_3$$
$$IR_e = I(R_1 + R_2 + R_3)$$

Dividing by I on both sides

$$R_e = R_1 + R_2 + R_3 \qquad (2.5)$$

Resistors in parallel

Figure 2.6 shows a typical parallel circuit using three resistors. In each branch circuit is connected an ammeter, A_1, A_2 and A_3 and there is also an ammeter connected to measure the supply current. Applying Kirchhoff's first law to the supply current, then:

$$I = I_1 + I_2 + I_3$$

but according to Ohm's law

$$I = V/R_e$$

For each part of the circuit

$$E/R_e = E/R + E/R + E/R$$
$$= E(1/R_1 + 1/R_2 + 1/R_3)$$

Dividing by E on both sides

$$1/R_e = 1/R_1 + 1/R_2 + 1/R_3 \quad (2.6)$$

The point to remember is that $1/R_e$ is the reciprocal of resistance and once this is found it needs to be inverted to $R_e/1$, the equivalent circuit resistance. In a parallel circuit, the value of R_e will always be less than the smallest resistive component in the circuit.

If a parallel circuit is dealing with only two resistors, then the method to use is the *product/sum*, i.e.

$$R_e = (R_1 \times R_2)/(R_1 + R_2)$$

Where a circuit comprises series and parallel connections, treat each parallel circuit separately and find its equivalent resistance. The combination can then be treated as a series circuit adding the groups together. It is important not to get confused with E and V, remembering that V is a p.d. and will be less than E the e.m.f. Consider the following examples:

Example 1

Three resistors, $1\,\Omega$, $2\,\Omega$ and $3\,\Omega$ are connected in series across a $12\,V$ battery. Find the equivalent circuit resistance, the current taken from the supply and the p.d. across each resistor.

Solution

$$R_e = R_1 + R_2 + R_3$$
$$= 1 + 2 + 3$$
$$= \underline{6\,\Omega}$$
$$I = E/R_e$$
$$= 12/6$$
$$= \underline{2\,A}$$

Across R_1, p.d. is $V = IR_1 = 2 \times 1 = \underline{2\,V}$
Across R_2, p.d. is $V = IR_2 = 2 \times 2 = \underline{4\,V}$
Across R_3, p.d. is $V = IR_3 = 2 \times 3 = \underline{6\,V}$
Thus $E = IR_1 + IR_2 + IR_3$

Example 2

Consider the three resistors in Example 1 now connected in parallel across the same supply source. Find the equivalent circuit resistance, the supply current and the branch currents in each resistor.

Solution

$$1/R_e = 1/R_1 + 1/R_2 + 1/R_3$$
$$= 1 + 0.5 + 0.333$$
$$= 1.833$$
$$R_e = \underline{0.545\,\Omega}$$
$$I = E/R_e$$
$$= 12/0.545$$
$$= \underline{22\,A}$$

Current in R_1 is $I_1 = E/R_1 = 12/1 = \underline{12\,A}$
Current in R_2 is $I_2 = E/R_2 = 12/2 = \underline{6\,A}$
Current in R_3 is $I_3 = E/R_3 = 12/3 = \underline{4\,A}$
Thus $I = I_1 + I_2 + I_3$

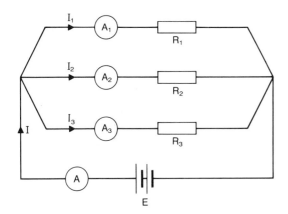

Fig. 2.6 Parallel circuit (common voltage).

It should be noted that both these circuits have identical components, yet the parallel circuit equivalent resistance is eleven times smaller than the series circuit. In real terms the parallel circuit will dissipate eleven times more heat. Resistors in series create voltage drops and their main purpose is to block current flow.

Example 3

Two 12 Ω resistors are connected in parallel across a 12 V battery. What is their equivalent resistance and the current taken from the battery?

Solution

$$R_e = (R_1 \times R_2)/(R_1 + R_2)$$
$$= (12 \times 12)/(12 + 12)$$
$$= \underline{6\ \Omega}$$
$$I = E/R_e$$
$$= 12/6 = \underline{2\ A}$$

One should note that where two or more *identical* resistors are connected in parallel, the equivalent resistance is divided by the number of resistors. For example five 20 Ω resistors would be equivalent to a resistance of 4 Ω.

Example 4

With reference to Figure 2.7, determine the current flowing in each part of the circuit.

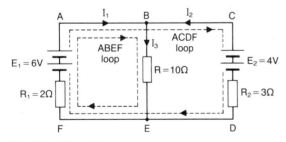

Fig. 2.7 Closed loop circuit.

Solution

The arrangement of this circuit is a little more complex than that already dealt with above and it requires the formation of simultaneous equations to solve the current values in different parts of the circuit. The secret in solving the problem is to make assumptions of current direction (we'll assume clockwise) and follow the rules of algebra. Applying Kirchhoff's second law to the loop ABEF, noting that $I_3 = I_1 + I_2$

Hence
$$E_1 = I_1 R + I_3 R$$
$$6 = 2I_1 + 10(I_1 + I_2)$$
$$6 = 12I_1 + 10I_2 \tag{1}$$

similarly for loop ACDF
$$E_1 - E_2 = I_1 R - I_2 R$$
$$6 - 4 = 2I_1 - 3I_2$$
$$2 = 2I_1 - 3I_2 \tag{2}$$

If we now multiply equation (2) by 6 and then subtract it from equation (1) it will reveal the value of I_2. Note: When subtracting, change signs and add (see (1) below).

Thus
$$12 = \quad 12I_1 - 18I_2 \tag{3}$$
$$-6 = -12I_1 - 10I_2 \tag{1}$$
$$\overline{\quad 6 = \qquad - 28I_2 \quad}$$
therefore $I_2 = -(6/28)$
$$= \underline{-0.214\ A}$$

This shows that I_1 is flowing against I_2 and makes sense since the battery voltage of E_1 is greater than E_2. The current I_2 can now be substituted in equation (2) and I_1 solved.

Thus:
$$2 = 2I_1 - 3I_2$$
$$= 2I_1 - 3(-0.214)$$
$$= 2I_1 + 0.642$$
therefore $I_1 = 1.358/2$
$$= \underline{0.679\ A}$$
Since $\quad I_3 = I_1 + I_2$
then $\quad I_1 = 0.679 - 0.214 = \underline{0.465A}$

Example 5

A more practical example of using Kirchhoff's laws is to consider a typical domestic 240 V, 30 A ring final circuit as shown in Figure 2.8. Socket outlets in use are shown at points B, C and D. Given the resistance of each section to be: AB = 0.09 Ω, BC = 0.07 Ω, CD = 0.06 Ω and DA = 0.08 Ω, find (a) the current in each section of the ring and its direction, and (b) the voltages at each load.

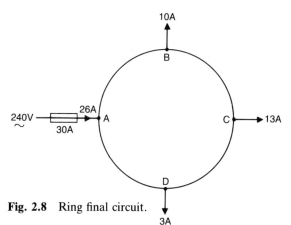

Fig. 2.8 Ring final circuit.

Solution

One point to mention in Kirchhoff's second law is that: *in any closed circuit, the sum of all the potential drops is zero* (i.e. $\Sigma IR = 0$). This can be written as: $0 = IR_1 + IR_2 + IR_3 + \ldots IR_4$. Let the current in each section of Figure 2.8 be as follows, taking a clockwise direction and subtracting the load points:

Between AB it is I A
Between BC it is $I - 10$ A
Between CD it is $I - 23$ A
Between DA it is $I - 26$ A

The section resistances are as follows:

Between AB, $R_1 = 0.09\ \Omega$
Between BC, $R_2 = 0.07\ \Omega$
Between CD, $R_3 = 0.06\ \Omega$
Between DA, $R_4 = 0.08\ \Omega$

Inserting these values in the formula:

$$0 = IR_1 + IR_2 + IR_3 + IR_4$$
$$0 = 0.09I + 0.07(I - 10) + 0.06$$
$$(I-23) + 0.08(I - 26)$$
$$0 = 0.09I + 0.07I - 0.7 + 0.06I -$$
$$1.38 + 0.08I - 2.08$$
$$0 = 0.3I - 4.16$$
$$4.16 = 0.3I$$

therefore $I = 4.16/0.3 = \underline{13.87\ A}$

The current in each section is as follows:

From AB it is $I = \underline{13.87\ A}$
From BC it is $13.87 - 10 = \underline{3.87\ A}$
From CD it is $13.87 - 23 = \underline{-9.13\ A}$
From DC it is $13.87 - 26 = \underline{-12.13\ A}$

The last two sections show that current is flowing in an anticlockwise direction. Note that the total current is 26 A.

The voltage at B is 240 V $- (13.87 \times 0.09)$
$= \underline{238.75\ V}$
The voltage at C is 238.75 V $- (3.87 \times 0.07)$
$= \underline{238.48\ V}$
The voltage at D is 240 V $- (12.13 \times 0.08) = \underline{239\ V}$
Note: Sum of volt drops (ΣIR)
$= 1.248 + 0.271 - 0.971 - 0.548 = 0$

Example 6

Figure 2.9 shows a radial circuit with loads being taken off at various points. Determine the current in each section AB, BC, CD and DE and the voltages across the respective loads.

Assume the combined phase and neutral conductor resistances in each section are the same, i.e. 0.02 Ω.

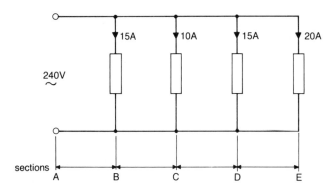

Fig. 2.9 Radial final circuit.

Solution

The current in each section is:
Section AB $I = 60$ A
Section BC $I = 60 - 15 = 45$ A
Section CD $I = 45 - 10 = 35$ A
Section DE $I = 35 - 15 = 20$ A
The voltage drop in each section is:
Section AB $V = IR = 60 \times 0.02 = 1.2$ V
Section BC $V = IR = 45 \times 0.02 = 0.9$ V
Section CD $V = IR = 35 \times 0.02 = 0.7$ V
Section DE $V = IR = 20 \times 0.02 = 0.4$ V
Since the supply voltage is 240 V, then the voltages across the loads are:

Section AB $V = 240.0 - 1.2 = 238.8$ V
Section BC $V = 238.8 - 0.9 = 237.9$ V
Section CD $V = 237.9 - 0.7 = 237.2$ V
Section DE $V = 237.2 - 0.4 = 236.8$ V

It should be noted that voltage drop leads to electrical equipment not working efficiently. Heating appliances may not reach their correct operating temperatures and discharge lamps may not light properly. The cause is often due to circuit conductors being either too small or too long or because the temperature of circuit conductors has increased abnormally. To safeguard against voltage drop problems, the IEE Wiring Regulation (Reg. 522–8) refers to a 2.5% limit placed on the declared supply voltage. This limit is taken from the origin of the circuit to any other point in the circuit. In the example shown above, this voltage drop limit is $2.5/100 \times 240 = 6$ V. The remote load should not have a supply to it less than 240 V $-$ 6 V $= 234$ V.

Energy and Power

A definition for both these terms has already been given in the previous chapter. Energy, being the capacity for doing work, exists in various forms. For example, *potential energy* is a form of energy which a body possesses by virtue of its position, such as a coiled spring. Another form is *kinetic energy* which a body possesses by virtue of its motion, such as when the spring is released. This transformation of energy obeys the common law of conservation which states that energy can neither be created nor destroyed but only changed from one form to another. The generation of electricity is a good example of energy conversion. In a power station, a fossil fuel such as coal or oil is burned to produce heat. This heat is used to boil water which is then converted into high-pressure steam to drive large turbines. These turbines are mechanically coupled to a.c. generators which produce the electricity required. The conversion cycle has actually passed through four main stages, namely, chemical energy, heat energy, mechanical energy and electrical energy.

All these forms of energy have the same basic unit of measurement which is called the *joule* (J).

In our basic electric circuit, the e.m.f. source is measured in terms of the number of joules of work

necessary to move one coulomb of electricity around the circuit. This can be expressed as:

$$E = W/Q \qquad (2.7)$$
where E is the e.m.f. in volts (V)
W is the energy in joules (J)
Q is the quantity of electricity in coulombs (C)

By transposition of formula (2.7),
then $W = EQ$

But we have already seen in (2.1)
that $Q = It$
therefore $W = EIt$ (2.8)

Power (P) is the rate of doing work such as when one joule of work is done in one second. It is the ratio of energy and time and its unit is the *watt*, named after James Watt (1736–1819). The ratio can be expressed as:

$$P = W/t \qquad (2.9)$$
by re-arranging the formula: $W = Pt$
but in formula (2.8) $W = EIt$
By combining these two formulae, we obtain for W:
$$Pt = EIt$$
Thus $P = EI$ (2.10)
This is the total power dissipated by the electric circuit. When, however, a current flows through an electrical component such as a resistor, heat is dissipated between the points of its connection. The power in this part of the circuit involves the current and the potential difference (V) and not the electromotive force (E) and expression (2.10) can be written as:

$$P = VI \qquad (2.11)$$

There are several ways of expressing power. For example, in the formula for
Ohm's law (2.2) $I = V/R$
and substituting I in (2.11) $P = VV/R$
                                              $P = V^2/R$ (2.12)
Also, since $V = IR$, then $P = IRI$
                                              $= I^2R$ (2.13)

The instrument which records energy is called an *integrating meter* or *energy meter* and the instrument which measures power is called a *wattmeter*. They are not to be confused, the former involves recording time whereas the latter is a measuring instrument only.

In the process of reading a consumer's energy meter, one will come across dial type meters as well as digital types which are a later development and gaining widespread use, particularly in conjunction with Economy 7 metering arrangements. The usual practice in reading an energy meter is to subtract the first reading from the second reading as was described in Volume 1, *Installation Theory & Regs*: Sometimes the energy meter can be used to find the rating of a piece of equipment by using a meter's disc constant such as 240 revolutions per kilowatt hour or every unit. From expression (2.9) $P = W/t$ the rating of the equipment is:

P = No. of revs/(time taken × 240)

For example, if the equipment were timed for 5 min (0.083 h) and 40 revolutions were counted as the disc rotated, then:

$$P = \frac{40}{0.083 \times 240}$$
$$= 2 \text{ kW}$$

Example 1

Two resistors of 12 Ω and 28 Ω respectively are connected in series and then in parallel across a 200 V supply. Which mode of connection will produce the most heat?

Solution

In series $R = 40 \text{ Ω}$

$\quad I = V/R = 200/40 = 5 \text{ A}$

therefore $P = VI = 200 \times 5 = \underline{1\,000 \text{ W}}$

In parallel $R = 8.4 \text{ Ω}$

$\quad I = V/R = 200/8.4 = 23.81 \text{ A}$

$\quad P = VI = 200 \times 23.81 = \underline{4\,762 \text{ W}}$

Example 2

The estimated daily loading of a consumer's final circuit is as follows:

Lighting	0.5 kW for 6 hours
Water heating	3.0 kW for 2 hours
Ring circuit	3.5 kW for 3 hours
Cooking	5.0 kW for 2 hours

Determine the energy used daily, in (a) joules and (b) kWh

Solution

(a) $W = Pt$

$\quad W = [(500 \times 6) + (3000 \times 2) + (3500 \times 3)$
$\quad\quad + (5000 \times 2)] \times 3600$
$\quad\quad \underline{106.2 \text{ MJ}}$

(b) $W = Pt$

$\quad (0.5 \times 6) + (3 \times 2) + (3.5 \times 3) + (5 \times 2)$
$\quad \underline{29.5 \text{ kWh}}$

Check: Since 1 kWh = 3.6 MJ,
then 29.5 kWh = 106.2 MJ
Note: If the cost of energy is 5.2 p/unit (kWh) then the daily charge is 29.5 × 5.2 = 153.4p or £1.53

Example 3

An electric kettle is rated at 3 kW/240 V. Find the cost of using the kettle on 28 occasions per week if it takes two minutes to boil. Take one unit of electricity to cost 5.8p.

Solution

Since $W = Pt = \text{kWh}$

then $W = 3 \times 2/60 \times 28 = 2.8 \text{ kWh}$

$\quad$ therefore cost/week = 2.8 × 5.8 = $\underline{16p}$

Note: Over one quarter (13 weeks), the cost would be £2.11.

Voltage sources

At this stage, it has been seen that current flow is a result of some form of electromotive force. In practice, a distinction is made between two main sources of supply, namely direct current (d.c.) and alternating current (a.c.). The former can be created chemically by cells or batteries or even from d.c. generators such as dynamo, whereas the latter is created by a.c. generators known as alternators. Both types of a.c. and d.c. generator use the principles of electromagnetic induction and it is the a.c. source which is used for public supplies into consumers' premises. It can be rectified to d.c. quite easily and it can also be transformed to higher or lower voltages using transformers.

Magnetic effects of current

Hans Christian Oersted (1787–1851) discovered that
by passing a direct current through a copper wire, a
compass needle held at the side of the wire, swung to
one side. When he reversed the current direction, the
compass needle swung to the opposite side. Oersted
had demonstrated that an electric current has a
magnetic effect.

The current-carrying conductor produces a
magnetic field which can be made visible by using
iron filings sprinkled on a flat card with the conductor
passing through the centre of it. By placing the
compass needle on the card the direction of the field
can be traced (see Figure 2.10). It will be seen that
these lines of force are concentric in shape. If current
is flowing inwards or down through the card, the
magnetic lines of force follow a clockwise direction
and if current is flowing outwards or up through the
card, the magnetic lines follow an anticlockwise
direction. This can be remembered by applying the
cross and dot rule as shown in Figure 2.11. It gives
both current direction and magnetic field direction.

Another way of remembering this is by using the
right-hand grip rule (see Figure 2.12) whereby the
curl of one's fingers represents the direction of
current and the thumb points in the direction of the
magnetic field. This rule is particularly useful in
determining the field pattern set up around a
solenoid, see Figure 2.13. Further examples using the
cross and dot notation are given in Figure 2.14. It is
worth mentioning that the existence of these
magnetic flux lines obey certain rules, for example,
they each form a closed loop and are always trying to
contract especially when stretched, and they never
intersect each other. Moreover, when they are parallel
and in the same direction they repel one another.

It was Michael Faraday (1791–1867) who
discovered the principles of *eletromagnetic induction*.
He found that when a bar magnet was moved
towards a stationary coil, the magnet's field created
electricity in the coil. This is shown in Figure 2.15. In
moving the bar magnet away from the coil, the
electricity flowed in the opposite direction. This
principle is the same if moving the coil instead of
moving the magnet.

Figure 2.16(a) shows a short piece of copper
conductor being moved inside a stationary permanent
magnet. The copper wire is connected to a sensitive

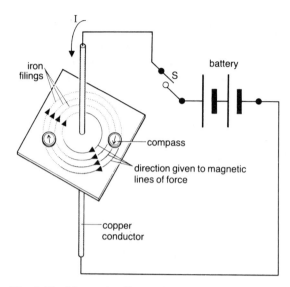

Fig. 2.10   Magnetic effects of current.

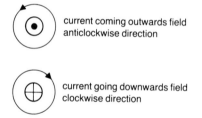

Fig. 2.11   Dot and cross rule.

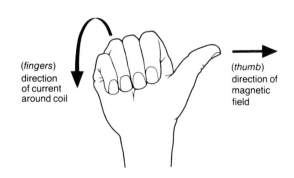

Fig. 2.12   Grip rule.

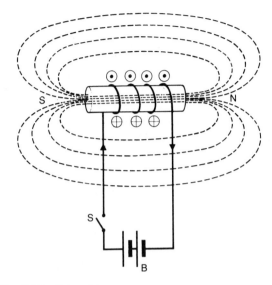

Fig. 2.13   Solenoid.

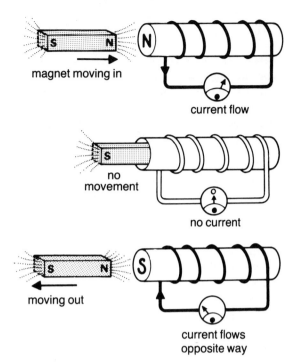

Fig. 2.15   Electromagnetic induction.

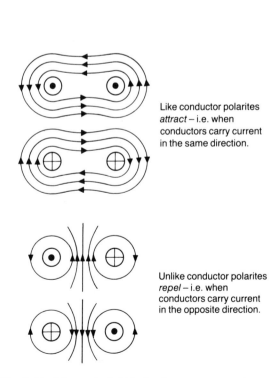

Like conductor polarites *attract* – i.e. when conductors carry current in the same direction.

Unlike conductor polarites *repel* – i.e. when conductors carry current in the opposite direction.

Fig. 2.14   Magnetic field behaviour.

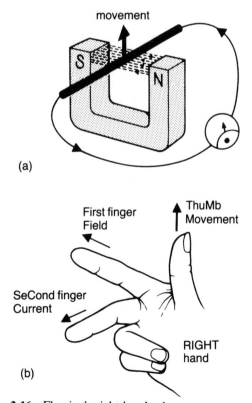

Fig. 2.16   Fleming's right-hand rule.

galvanometer. The galvanometer needle will deflect in one direction and then in the opposite direction when the wire is passed in and out of the field. The faster this is done, the greater will the needle deflect. The induced e.m.f. is equal to the average rate of cutting the magnetic flux.

Thus $E = \Phi/t$ $\qquad$ (2.14)

where E is the e.m.f. induced in volts (V)

$\qquad$ Φ is the magnetic flux in webers (Wb)

$\qquad$ t is the time in seconds (s)

There are three factors which alter the strength of an induced e.m.f. namely, (i) the strength of the magnetic flux density (B) between the magnetic poles (ii) the length (l) of conductor in the magnetic field and (iii) the velocity (v) or speed of the conductor passing through the magnetic field.

Hence $E = Blv$ $\qquad$ (2.15)

In practice, there are two methods of determining the direction of induced e.m.f., one is by *Fleming's right-hand rule* and the other is by *Lenz's Law*.

John Fleming (1849–1945) introduced a right-hand rule for generators and a left-hand rule for motors and Figure 2.16(b) shows how the fingers and thumb of the right hand are kept at right angles to indicate e.m.f., magnetic flux and motion. Students ought to apply this rule to Figures 2.17 and 2.18 and they must accept the convention used that magnetic lines of force emanate from a north-seeking pole and enter a south-seeking pole.

It was Heinrich Lenz (1804–65) who discovered that when a circuit and a magnetic field move relatively to each other, the direction of induced e.m.f. is always such that it tends to set up a current opposing the motion or the change of flux responsible for inducing the e.m.f. This is demonstrated in Figure 2.15 where it will be seen that by moving the permanent magnet closer to the coil a similar pole is created at the end of the coil. This tells us that the current is opposing the movement of the magnet by trying to repel it. Moving the magnet away makes the current in the coil move the other way to produce an opposite pole which again opposes the movement of the magnet by trying to attract it.

Generation of a direct current

Figure 2.17(a) shows the simple construction of a single-loop d.c. generator. It has a part on it called a *commutator* so that it can draw a unidirectional (d.c.) e.m.f. from the rotating loop. This will be achieved by the brushes resting on the commutator. When the loop is rotated there will be maximum e.m.f. induced in the position shown, i.e. under the main poles. As the loop moves further around, less e.m.f. will be induced and no e.m.f. will be induced when the loop is travelling horizontal to the main field path. It is at this point that the commutator segments change and the induced e.m.f. begins to rise again as the loop moves toward the middle of the main poles again. The external current will rise to a maximum when this happens and then fall to zero on each half cycle of 180°. This will be repeated as shown by the graph in Figure 2.17(b). The more commutator loops and more segments there are, the smoother will be the d.c. output.

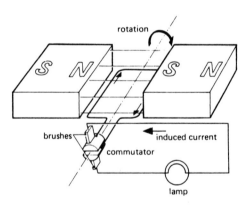

(a) Single loop generator showing commutator

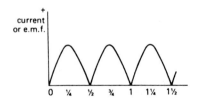

(b) Output waveshape

Fig. 2.17 Generation of direct current.

Generation of an alternating current

Figure 2.18(a) shows how a single-loop a.c. generator operates. Unlike the d.c. generator the loop is attached to slip rings but again brushes are used to supply an external circuit. When the loop is rotated, maximum flux cutting occurs underneath the main poles (positions 3 and 7) i.e. at right angles to the field. In positions 1 and 5 the loop is travelling horizontally to the field and no induced e.m.f. occurs. Since the slip rings are not segmented the output supply will be cyclic over one complete revolution and be of sinusoidal wave shape. Positions 3 and 7 are the maximum values of induced e.m.f. in one side of the loop and it should be noted that one side of the loop changes its polarity at position 5, i.e. it generates positive and negative values (see Figure 2.18(b)).

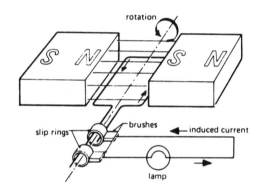

(a) a.c. generator

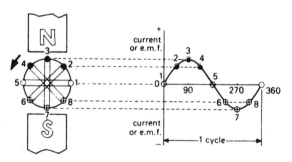

(b) Output waveshape

Fig. 2.18 Generation of alternating current.

Force on a conductor

It has already been seen from Figure 2.14, using the cross and dot notation, that a force of attraction or repulsion is created between two single conductors when carrying current. Such force is a direct result of the interaction between the two magnetic fields.

However, it can be seen from Figure 2.19 that a force is also experienced on a single current-carrying conductor when it is situated inside a permanent magnetic field. The conductor's own magnetic field interacts with the permanent magnet's field, and the combined effect is strong above the conductor and weak below it. This results in the conductor being ejected at right angles to the main field. It is Fleming's left-hand rule (see Figure 2.20) which is used to interpret the direction of this force and is a useful rule to know, when trying to understand the operation of some measuring instruments and direct current motors. An example of this is shown in Figure 4.11 where it will be seen that a force is exerted on both sides of a single loop conductor resulting in an anticlockwise motion. The magnitude of this force (F) in newtons (N) is found to be proportional to the strength of the magnetic flux or magnetic flux density (B) in teslas (T), the current (I) in amperes (A) and active length of conductor (l) in metres (m) within the field. This can be expressed as:

$$F = BIl \tag{2.16}$$

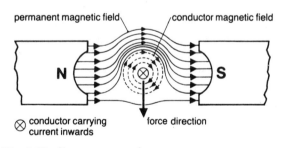

Fig. 2.19 Force on a conductor.

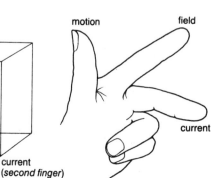

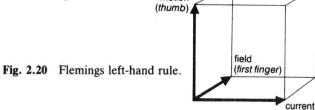

Fig. 2.20 Flemings left-hand rule.

Example

A copper conductor situated at right angles to a uniform magnetic field of flux density 5 T, carries a current of 20 A. If the field is 25 mm in the direction of the conductor, calculate the force produced.

Solution

$$F = 5 \times 20 \times 25 \times 10^{-3}$$
$$= \underline{2.5 \text{ N}}$$

It was shown in Figure 2.13 that a solenoid is capable of creating a magnetic field and this is the basic principle of all electromagnets. An electromagnet's construction is not merely an open coil like the solenoid, it is an insulated winding on a soft iron core and the magnetic flux it produces is created by *magnetomotive force* (abbreviated m.m.f.). This m.m.f. depends on the current (*I*) flowing in the winding and the number of conductor turns (*N*) wound around it.

The symbol for m.m.f. is also *F* and must not be confused with the same symbol for force. If the solenoid (Figure 2.13) carried a current of 5 A it would produce an m.m.f. of 20 *ampere turns* – since it is wound with four turns. Electromagnets have many uses in heavy industry such as lifting magnets for scrap metal, (see Figure 2.21); creating rotation in electric motors; operating contactors and relays; and even the operation of the simple bell circuit shown in Figure 2.22. On pressing the bell push, current completes the circuit and the soft iron armature becomes attracted to the electromagnet causing the bell to ring. The movement of the armature breaks the electric circuit at the contact screw (point X) and the magnetic field ceases to exist. Keeping the push depressed allows the bell to continually ring.

Inductance

Any circuit in which a change of current is accompanied by a change of magnetic flux resulting in an induced e.m.f. is said to be *inductive*. The unit of inductance (*L*) is called the *henry* (H) and is named after Joseph Henry (1797–1878). He found that a circuit has an inductance of 1 henry if an e.m.f. (*e*) of 1 volt is induced in it by a current

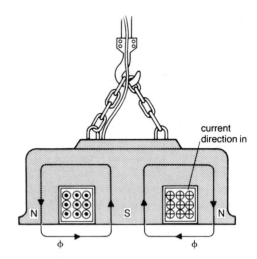

Fig. 2.21   Lifting magnet.

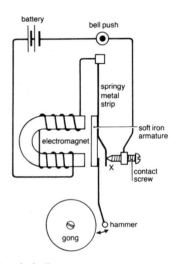

Fig. 2.22   Electric bell.

changing at a rate of 1 ampere per second (*I/t*). This can be stated as:

$$L = e/(I/t) \qquad (2.17)$$

The average rate of change of current (*I/t*) is expressed as the increase in current divided by the time for the current to increase, i.e.

$$\frac{\text{final current } (i_2) - \text{initial current } (i_1)}{\text{time taken } (t)}$$

The following example shows how inductance can be determined in a coil.

Example

The e.m.f. induced in a coil is 240 V when the current through it changes from 1 A to 3 A in a time of 0.01 s. Calculate the inductance of the coil.

Solution

The average rate of change
of current
$$= (i_2 - i_1)/t$$
$$= (3 - 1)/0.01$$
$$= 200 \text{ A/s}$$
Therefore
$$L = e/(I/t)$$
$$= 240/200 = \underline{1.2 \text{ H}}$$

The expression (2.17) can be re-arranged to find the e.m.f. induced at any instant. In this case: $e = -L \times$ **rate of increase of current.** The minus sign indicates that the self-induced e.m.f. opposes the increase in current. It is worth pointing out that when current is rapidly switched off in a circuit containing inductance, such as an electromagnet or coil, the collapsing current and magnetic field give rise to a large induced e.m.f. and this in turn produces a momentary arc at the switch contacts. This could cause damage to the contacts and to the circuit cable insulation. This effect is often visible in the bell circuit previously described but in a.c. circuits care in the selection of correctly rated switches is required, particularly for discharge lamp circuits. Consider the following example:

Example

An inductor in a circuit has an inductance of 1.2 H and carries a current of 15 A. What is the induced e.m.f. created if the circuit is switched off in 0.01 s?

Solution

$$e = -L \times \text{rate of change of current}$$
$$= -1.2 \times (0 - 15)/0.01$$
$$= \underline{1800 \text{ V}}$$

The effects of rapidly switching on and off and inductive circuit is shown in Figure 2.23. The energy stored in the circuit is *magnetic energy* and is given by the formula: $W = \frac{1}{2}LI^2$ joules.

It should have been noted earlier in the resistive component, that energy used is converted into heat

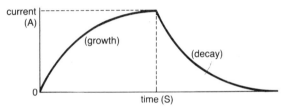

Fig. 2.23 Charge and discharge currents of an inductor.

energy and from the formula (2.8), since $W = EIt$ and $E = IR$, then $W = I^2Rt$ joules.

Capacitance

Capacitance is the property of a capacitor to store charge when its plates are at a different potential. Figure 2.24 shows a simple parallel plate capacitor. It consists of two conducting surfaces arranged in close proximity to each other and separated by an insulating material known as the *dielectric*. The area of its plates, the spacing between them and the type of material used as the dielectric all influence the capacitor's capacitance. In general, the capacitance of a capacitor is proportional to the area of its plates and inversely proportional to the distance between them. Its capacitance is also dependent upon the nature of the dielectric used; for example, the capacitance would be increased if paper, glass or mica were used rather than air. It is the type and thickness of dielectric material which also governs the maximum working voltage of a capacitor, since too high a voltage will cause it to breakdown.

In theory, when the switch in Figure 2.24(a) is closed, electrons on the X-plate of the capacitor are attracted to the positive potential of the battery and they end up as surplus electrons of the Y-plate. The p.d. rising on the plates of the capacitor oppose the battery e.m.f. as the capacitor becomes fully charged. When this point is reached no electron movement or current occurs. The capacitor's plates now have equal and opposite charges and it acts like a reservoir of electrical energy, being charged to the same potential as the battery. Figure 2.25 shows a capacitor's charge and discharge characteristics. It is normal practice to discharge a charged capacitor through some form of resistive component as a safety precaution.

The unit of capacitance (C) is called the *farad* (F) and for this a capacitor requires a potential difference

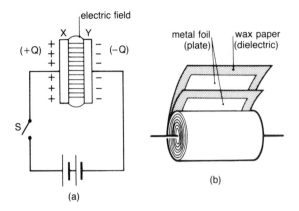

Fig. 2.24 (a) Parallel plate capacitor (b) metal foil capacitor.

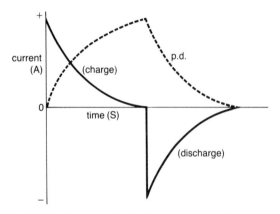

Fig. 2.25 Charge and discharge currents of a capacitor.

(*V*) of 1 V to maintain a charge (*Q*) of 1 coulomb. Thus the relationship between charge, p.d. and capacitance is:

$$C = Q/V \qquad (2.18)$$

A capacitance of 1 farad is abnormally large and in practice it is more usual to find the sub-unit *microfarad* (μF) being used.
Consider the following example:

Example

What is the capacitance of a parallel-plate capacitor holding a charge of 50 mC on its plates when it is connected to a supply of 500 V?

Solution

Since $C = Q/V$
then $C = 0.05/500$
           $= 0.000\ 1$ F
or $C = \underline{100\ \mu F}$

Capacitor connections

Capacitors are often connected in series or parallel in the same way as connecting resistor components. However, in series, the capacitance is not increased as is resistance in a series circuit. Similarly, in parallel, the plates form one larger plate and unlike parallel resistors whereby the equivalent resistance decreases, in parallel capacitors, the capacitance is increased.

Capacitors in series

Figure 2.26 shows a typical series circuit where C_1, C_2 and C_3 are three capacitors and V_1, V_2 and V_3 are three voltmeters connected across each component. Since the sum of the voltmeter readings (p.d.s) is equal to the applied battery voltage, i.e.
$V = V_1 + V_2 + V_3$

From 2.18 above,
since $C = Q/V$ and $V = Q/C_e$
then $Q/C_e = Q/C_1 + Q/C_2 + Q/C_3$
                              $= Q(1/C + 1/C + 1/C)$
Dividing by Q on both
sides $1/C_e = 1/C_1 + 1/C_2 + 1/C_3$ (2.19)
The equivalent circuit capacitance is found by inverting $1/C_e$.

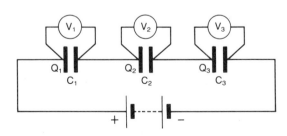

Fig. 2.26 Series connected capacitors.

Capacitors in parallel

Figure 2.27 shows this arrangement with the same battery voltage across each component. Since the total charge is $Q = Q_1 + Q_2 + Q_3$
From 2.18 above, Q can be expressed as $Q = C_eV$

Thus
$$C_eV = C_1V + C_2V + C_3V$$
$$= V(C_1 + C_2 + C_3)$$

Dividing by V on both sides
$$C_e = C_1 + C_2 + C_3 \quad (2.20)$$

Consider the following example:

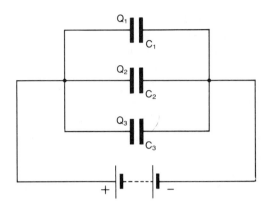

Fig. 2.27 Parallel connected capacitors.

Example

Three capacitors which have capacitances of 1 μF, 2 μF and 3 μF are connected across a 12 V battery. Determine the equivalent capacitance when connected in (a) parallel and (b) series. In the parallel connection, determine the charge on each capacitor and the total charge, and in series connection the total charge and the p.d. across each capacitor.

Solution

In parallel
$$C_e = C_1 + C_2 + C_3$$
$$= 1 + 2 + 3 = 6 \text{ μF}$$

Since
$$Q = VC$$

The charge on the 1 μF capacitor is
$$Q_1 = 1 \times 12 = 12 \text{ C}$$

The charge on the 2 μF capacitor is
$$Q_2 = 2 \times 12 = 24 \text{ C}$$

The charge on the 3 μF capacitor is
$$Q_3 = 3 \times 12 = 36 \text{ C}$$

The total charge is
$$Q = Q_1 + Q_2 + Q_3$$
$$= 12 + 24 + 36 = 72 \text{ C}$$

or using equivalent circuit capacitance
$$Q = CV$$
$$= 6 \times 12 = 72 \text{ C}$$

In series
$$1/C_e = 1/C_1 + 1/C_2 + 1/C_3$$
$$= 1/1 + 1/2 + 1/3$$
$$= 1 + 0.5 + 0.333$$
$$= 1.833$$

therefore
$$C_e = 0.545 \text{ μF}$$

Since
$$Q = VC_e$$

then
$$Q = 12 \times 0.545$$
$$= 6.54 \text{ μC}$$

Since p.d. is given by the expression
$$V = Q/C$$

the p.d. across the 1 μF capacitor is
$$V = 6.541/1 = 6.54 \text{ V}$$

the p.d. across the 2 μF capacitor is
$$V = 6.54/2 = 3.27 \text{ V}$$

the p.d. across the 3 μF capacitor is
$$V = 6.54/3 = 2.18 \text{ V}$$

It will be noticed that because of the re-occurring decimal in the 3 μF capacitor the total p.d. only approximately equals the supply voltage.

Note: The total energy stored in the electric field of a capacitor is given by the formula $W = \frac{1}{2}CV^2$ joules. This energy should be compared with the energy used by a resistive component and also by an inductive component.

Chemical effects of current

The effect of an electric current flowing through a liquid is to try to separate the liquid into its chemical parts. This splitting up of the liquid's chemical compounds is a process called *electrolysis*. Faraday discovered this effect when he passed a current between two copper plates which were immersed in a copper sulphate solution. Any liquid solution like this substance which can conduct electricity is referred to as an electrolyte. Faraday found that copper was deposited on the negative cathode plate. From his experiments, he deduced two laws:

(1) The mass (m) of a substance deposited in an electrolytic cell is proportional to the current (I) and to the time (t) for which it passes;
(2) The masses of different substances deposited or liberated by the same quantity of electricity are proportional to the various chemical equivalents of the substances (z).

These two laws result in the formula $m = Itz$ and since $Q = It$ (see equation 2.1) then $m = Qz$.
Re-arranging this formula
$$z = m/Q \text{ gram/coulomb} \qquad (2.21)$$

Figure 2.28 shows the arrangement of a copper voltameter used for determining the electrochemical equivalents of metals. Copper is found to be 0.3294 mg/C whereas silver is 1.1182 mg/C.

The process in which metals are made to deposit on the cathode electrode is called *electro-plating*. The idea is to cover a base metal such as steel, with a very thin film of a much more expensive metal which is able to not only resist corrosion but also provide an attractive appearance such as gold, silver, nickel and chromium platings.

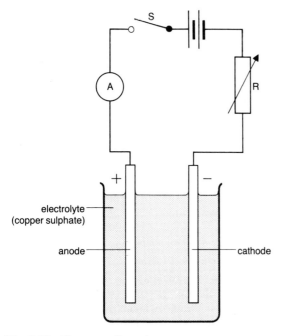

Fig. 2.28 Copper voltameter.

electrolyte
(copper sulphate)

anode

cathode

Electrolytic corrosion

Corrosion is basically an electrochemical reaction of a metal with its environment in which the metal reverts back to its natural oxide state. All metal surfaces create small electrical potentials because of differences brought about by flaws, impurities, etc. These potentials cause corrosion currents to flow in an oxidation process. Most students will no doubt be aware of the corrosion created by local action in a primary cell with the zinc electrode in contact with the electrolyte.

Precautions against corrosive substances or polluting substances are mentioned in Regs. 523–17 to Reg. 523–18 and also Appendix 10 of the *IEE Wiring Regulations*. The appendix mentions dissimilar metals liable to set up electrolytic actions and mentions contact between bare aluminium and any parts made of brass or other metal having a high copper content. Dampness and moisture play a very important part in the oxidation process and there are several ways to protect metals against this, such as painting or the use of inhibitors which act as a coating on the metal surface in an attempt to separate the metal from the corroding medium. With metal coatings, steel is often zinc coated in a process called *galvanizing* or it can be heated in a zinc dust process called *sheradizing*. A fuller description of protection against corrosion is given in Vol. 3 *Advance Work* by the same author.

Secondary cells

A primary cell is a cell which has no further use when its chemical composition is exhausted but a secondary cell is one that has the capability of being *re-charged* by passing current through it. The two most common types of secondary cell are (a) the *lead–acid cell* and (b) the *alkaline cell*.

Both these cells or batteries provide important commercial duties and, apart from daily routine use such as starting and running vehicles, they are used frequently as a back-up for standby emergency lighting as well as small power in places such as hospitals, cinemas, theatres, banks and other important buildings.

The basic lead–acid cell consists of two sets of plates immersed in an electrolyte. One type known as

Plante cells has a positive plate of pure lead and a negative lead plate which is a pasted grid containing lead oxides. The electrolyte used is pure sulphuric acid having a specific gravity of between 1.205 and 1.215 when the cell is fully charged. Figure 2.29 shows the construction of this cell which has a nominal voltage of 2 V per cell. The final discharge voltage should not be allowed to fall below 1.8 V.

There are several other types of lead–acid cell such as the *flat plate* and *tubular plate*, the difference between them being their positive plate arrangement, the type of container and their expected life cycle. For example, the Plante cells have a life cycle of about twenty years in which they provide 100% capacity. They are somewhat bulky for a given ampere-hour capacity and they are expensive. The tubular type is mechanically robust, less costly but has a shorter life cycle of between ten and twelve years. The flat-plate type is also less expensive than the Plante type and it is more compact, but like the tubular-plate type it suffers from having a shorter life cycle of about ten years.

These three types of lead–acid cell are all vented and designed to a capacity of several hundred ampere-hours. One modern form of lead–acid cell is the *sealed* type and is used where there is no risk of the electrolyte being spilled. They can be charged and discharged or stored in any position and whilst they are relatively cheap there are no maintenance requirements. Unfortunately, their life cycle is limited to seven years and their ampere-hour capacity is not as high as the other types mentioned. They also lose this capacity with age and storage.

Alkaline cells take the form of *nickel–iron* or *nickel–cadmium*. They can be sealed or vented types. The sealed types have the advantage that they are gas tight and require no maintenance. They have an indefinite shelf life even when discharged. However, they are relatively expensive and they have a low ampere-hour capacity (10 Ah) with a life expectancy of about seven years. The vented types have a much greater life expectancy of 20–25 years and provide a good performance over a wide temperature range. They can also be left in a discharged state without

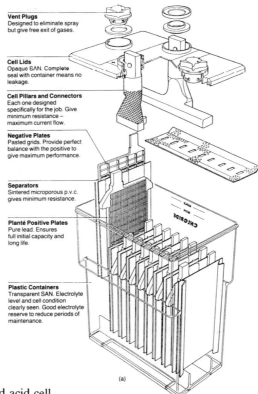

Vent Plugs
Designed to eliminate spray but give free exit of gases.

Cell Lids
Opaque SAN. Complete seal with container means no leakage.

Cell Pillars and Connectors
Each one designed specifically for the job. Give minimum resistance – maximum current flow.

Negative Plates
Pasted grids. Provide perfect balance with the positive to give maximum performance.

Separators
Sintered microporous p.v.c. gives minimum resistance.

Planté Positive Plates
Pure lead. Ensures full initial capacity and long life.

Plastic Containers
Transparent SAN. Electrolyte level and cell condition clearly seen. Good electrolyte reserve to reduce periods of maintenance.

(a)

Fig. 2.29 (a) Typical modern lead acid cell.

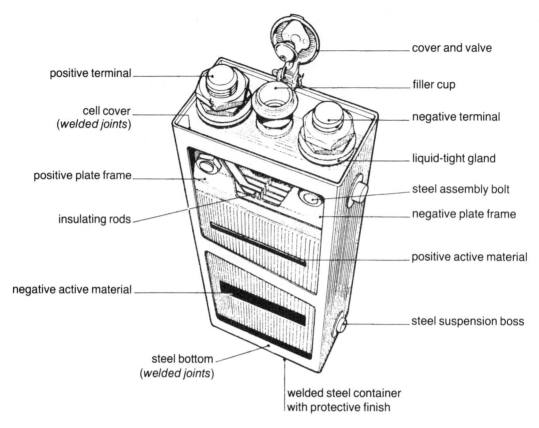

positive terminal

cover and valve

filler cup

cell cover
(*welded joints*)

negative terminal

liquid-tight gland

positive plate frame

steel assembly bolt

negative plate frame

insulating rods

positive active material

negative active material

steel suspension boss

steel bottom
(*welded joints*)

welded steel container
with protective finish

Fig. 2.29   (b) Nickel cadmium cell.

being damaged but again they are a little expensive in comparison with the cost of Plante cells. They unfortunately suffer a wide voltage drop from 1.4 V per cell down to 1.0 V per cell.

One type of nickel–cadmium cell is the *Alcad cell* (Figure 2.29(b)) and it uses nickel hydrate for its positive plate which is combined with pure graphite. Its negative plate consists of cadmium oxide combined with a special oxide or iron. The electrolyte is pure potassium hydroxide in distilled water and it serves only the purpose of conducting current between the plates. Only distilled water should be used for topping-up purposes. Its specific gravity is between 1.151 and 1.2 depending on the type and should not fall below 1.145. The nominal voltage per cell is 1.2 V. Smaller cells like the Cyclon type which are rechargeable provide 2 V per cell. They are sealed batteries with small ampere-hour capacities.

Charging and maintenance

There are several different methods of charging secondary cells. They are often divided into (i) *constant current* and (ii) *constant potential*. The first method may be split into a slow charge or fast charge but involves the use of a charging source capable of maintaining a constant current throughout the charge period. The latter method of charging involves the use of a constant-voltage source to the cells and the value of charge current will vary according to the state of the cells. The high-performance Plante cells use either method: the slow charge is sometimes referred to as *trickle charging*, i.e. a method of keeping the cells fully charged by passing a small current through them which will not cause gassing or allow the specific gravity to fall over a period of time. The constant-voltage method is often referred to as *float charging*, i.e. keeping the voltage applied to the cells at 2.25 V per cell and is the method used where continuous and variable loads exist. This method of charging has the advantage that it does not require manual attention.

Figure 2.30 shows a circuit diagram of a simple battery charging circuit using the constant-current method. The series variable resistor called a rheostat is adjusted to increase or decrease the charge current through the battery. It is important to use a direct-current source such as a rectifier unit and to make sure that its positive terminal is connected to the battery's positive terminal. A voltmeter can be connected across the battery to check its terminal voltage. The following examples serve to illustrate the two methods of battery charging. Internal resistance has been ignored.

Constant-current method

Three 2 V lead acid cells are to be re-charged from a 12 V d.c. supply source at a current of 4 A. When tested at the beginning of charge the terminal voltage of each cell was 1.9 V and at the end of the charging period it was 2.7 V. What is the value of the rheostat at the beginning and end of charge?

At the beginning of
charge the cells voltage $\quad E = 3 \times 1.9 = 5.7\,V$
The supply voltage
to the cells $\qquad\qquad V_s = 12\,V$
Since $\qquad\qquad\qquad V_s = E + IR$
then $\qquad\qquad\quad V_s - E = IR$
$\qquad\qquad\quad 12 - 5.7 = 4R$
and $\qquad\qquad\qquad\quad R = 6.3/4 = \underline{1.575\,\Omega}$
At the end of charge
the cells voltage $\qquad\quad E = 3 \times 2.7 = 8.1\,V$
since $\qquad\qquad\quad V_s - E = IR$
$\qquad\qquad\quad 12 - 8.1 = 4R$
$\qquad\qquad\qquad\quad = 3.9/4 = \underline{0.975\,\Omega}$

Constant-voltage method

Four 2 V lead acid cells are to be re-charged from a 12 V d.c. supply source. At the beginning of charge, each cell voltage is found to be 1.9 V and the charging current is 4 A. What is the final current at the end of charge if each cell voltage had increased to 2.7 V?

Solution

The constant voltage supplying the cells does not alter but the cells voltage increases in opposition to it and in doing so limits the current flowing.

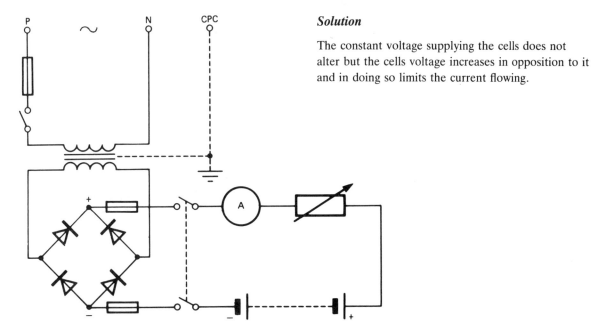

Fig. 2.30 Battery charging circuit.

At the beginning of
charge the cells voltage $E = 4 \times 1.9 = 7.6$ V
The supply voltage
to the cells $V_s = 12$ V
Since $V_s - E = IR$
             $12 - 7.6 = 4R$
                  $R = 4.4/4 = \underline{1.1 \ \Omega}$

This resistance is a fixed value in the charger unit.
At the end of charge
the cells voltage $E = 4 \times 2.7 = 10.8$ V
Since $V_s - E = IR$
then $12 - 10.8 = I1.1$
and $I = 1.2/1.1 = \underline{1.09}$ A

The state of charge of a lead-acid cell is directly
proportional to the specific gravity of its electrolyte
(i.e. the ratio of its relative density to that of water)
and this can be found by using a hydrometer as
shown in Figure 2.31. It consists of a weighted bulb
with a graduated slender stem and it floats vertically
in the electrolyte being tested. The apparatus will
expose a greater stem length with an electrolyte
having a higher density than one with a lower
density. In other words, the level to which the float
sinks, measures the specific gravity. In practice the
electrolyte of a lead-acid cell does not deteriorate
throughout its life, but the unsealed nickel–cadmium
types should have their specific gravity checked every
year since it does not vary with the state of charge
but falls gradually in service. Both lead–acid and
alcad types need occasional topping up with distilled
water and they give off hydrogen and oxygen during
quick charge. One should not attempt to inter-change
electrolytes of these cells.

From the point of view of safety, one should keep
a battery room well ventilated and the cells being
charged should be kept clean and dry. Cell
connections should be kept tight and covered with
petroleum jelly in order to protect against corrosion.
Cells should be re-charged as soon as possible after
discharge and records should be kept of each cell's
voltage. One should avoid metal objects falling across
terminals which could cause a spark and no naked
lights should be allowed in the room. When mixing
or handling electrolyte, personnel should wear
protective clothing such as goggles and rubber gloves.

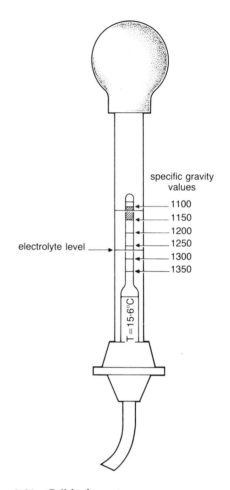

Fig. 2.31   Cell hydrometer.

Internal resistance (*r*)

The internal resistance of a secondary cell is made up
of the resistance of its plates, the connections and the
electrolyte. For large cells it is extremely low and for
any type of cell it is found by the expression:

$$r = (E-V)/I \tag{2.22}$$

where E is the cell's e.m.f.
       V is the cell's p.d.
       I is the circuit current

It should be mentioned that a cell's p.d. is less than
its e.m.f. when it is connected in circuit and this is
due to its internal resistance. For example, the open-
circuit voltage of a lead–acid cell is 2 V. When the

cell is discharging at the rate of 10 A, its terminal voltage falls to 1.9 V. What is the cell's internal resistance?

Using (2.22) $r = (2-1.9)/10 = 0.01\ \Omega$

Figure 2.32 shows how cells can be connected to give either a series or parallel arrangement or a combined series/parallel arrangement. In the case of the series arrangement, all the cell e.m.f.s are added together and all the internal resistances are added together. If there are nine cells and each cell has an e.m.f. of 2 V and internal resistance 0.01 Ω, then the battery would have an open circuit terminal voltage of 18 V and its internal resistance would be 0.09 Ω. In the parallel connection the battery open circuit voltage is maintained at 2 V but its internal resistance is now only one ninth of the original value, i.e. 0.0011 Ω. In the series/parallel arrangement, the battery voltage is 6 V and the internal resistance is now that of only one cell, i.e. 0.01 Ω.

Efficiency

The *ampere-hour capacity* has been used several times to describe a cell's capability but it really describes the ratio of the quantity of electricity taken out of the cell to the quantity of electricity required to be put back. It should be noted that a secondary cell is a *storage cell* and sometimes it is called a storage battery or accumulator. Since, however, *efficiency* usually expresses *output/input* then the cell's ampere-hour (Ah) efficiency can be expressed as:

Ah efficiency = Ah given on discharge/Ah required to charge

The ampere-hour efficiency is always stated for a given set of discharge conditions. In practice, for most lead–acid cells at the 10-hour rate of discharge this is about 90%. This means that an additional 10% needs to be put back into recharging the cell. One thing to note is that higher discharge currents give lower efficiencies. Consider the following example:

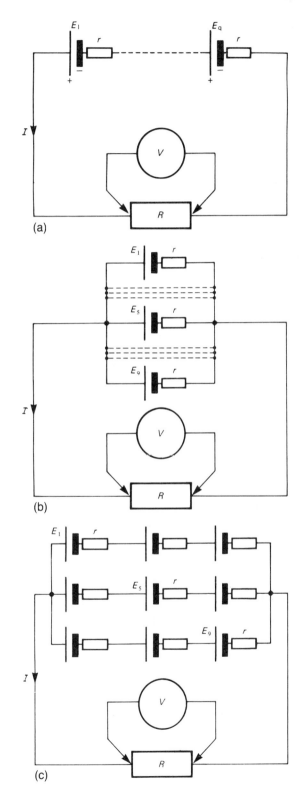

Fig. 2.32 Cell connections.

Example

A lead–acid cell is being discharged at the rate of 3 A for 15 hours. If its efficiency is 90% what is the ampere-hour charge required?

Solution

Discharge rate is 3 A × 15 h = 45 Ah
Since Ah efficiency = Ah discharge/Ah charge
then   Ah charge = Ah discharge/Ah efficiency
            = 45/0.9 = 50 Ah
Another useful efficiency is called the *watt-hour efficiency* and it expresses the ratio:

> Wh efficiency = watt-hours during discharge/watt-hours during charge

Example

A secondary battery is charged at 4 A for 10 hours and then discharged at 5 A for 6 hours. During charge the average battery voltage is 16 V while on discharge it is 12 V. Determine the battery's Ah efficiency and Wh efficiency.

Solution

Ah efficiency = Ah discharge/Ah charge
= (5 × 6)/(4 × 10) = 0.75 (75%)
Wh efficiency = Wh discharge/Wh charge
= (12 × 5 × 6)/(16 × 4 × 10) = 0.56 (56%)
Note: The Wh efficiency takes into consideration that the battery requires a higher voltage for charging.

Exercise 2

1   Three resistors of 8 Ω, 12 Ω and 24 Ω, respectively, are connected across a 220 V supply. Determine the equivalent resistance when they are connected in (a) series, (b) parallel. In each case find the power consumed.

2   Figure 2.33 shows the connection of four capacitors. Determine:
       a) the equivalent circuit capacitance
       b) the total charge
       c) the total energy stored.

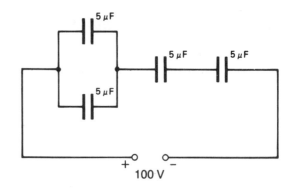

Fig. 2.33   Capacitor connections.

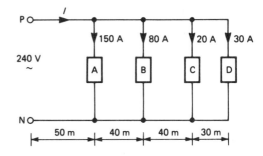

Fig. 2.34   Radial distributor.

3   With reference to Figure 2.34 showing a radial distribution system feeding various loads, determine:
       a) the supply current
       b) the voltage at each load
     Assume the cable is two-core and has a resistance of 0.1 Ω/1000 m.

4   A piece of copper wire 10 m long and 10 mm² in cross-sectional area carries a current of 5 A when it is connected to a 240 V supply. Determine the resistivity of the wire.

5   *a*) A conductor is 50 mm in length and lies at right-angles to a magnetic field strength 25 T. If it carries a current of 10 A calculate the force on the conductor.
      b) Give *two* practical uses of the above effect and briefly explain the interaction of the magnetic fields causing the conductor to move.

6   Briefly explain the difference between alternating current and direct current. State

what type of supply a transformer requires.

7   A coil has a resistance of 300 Ω at 0°C. If its resistance is found to increase to 330 Ω at a temperature of 25°C determine its temperature coefficient of resistance.

8   Calculate the average value of induced e.m.f. in an inductive coil of 0.7 H if the current through it increases from 2 A to 10 A in 0.04 s.

9   A battery consisting of nine primary cells is connected to an external resistance of 10 Ω. If each cell has an e.m.f. of 1.5 V and an internal resistance of 0.45 Ω, determine the circuit current and voltage drop across the 10 Ω resistor when the cells are connected in (a) series, (b) parallel, and (c) three sets in parallel, each set consisting of three cells in series.

10   Draw a neatly labelled wiring diagram of a simple battery charger circuit fed from a single-phase a.c. supply incorporating a double-wound transformer, bridge rectifier, ammeter and rheostat. Show also how a secondary battery is connected for charging purposes.

CHAPTER THREE

Alternating current circuits

After reading this chapter you will be able to:

1 Draw a.c. sinusoidal waveforms of voltage and current.

2 State the meaning of power factor and know its three conditions.

3 Perform calculations on RLC series and parallel circuits.

4 Draw phasor diagrams relating to RLC circuits.

5 Draw phasor diagrams relating to power circuits involving kW, kVA and kVAr.

6 State the difference between star and delta connections.

7 Perform load calculations in three-phase systems.

The sinewave

The generation of a.c. has already been mentioned briefly in Chapters 1 and 2, where it was seen that a rotating conductor situated in a magnetic field produced a voltage by electromagnetic induction. This induced or generated voltage takes the shape of a sinewave as was shown in Figure 1.10. The graph indicates that it not only reaches two maximum values, one in either direction, but it also passes through zero twice in each cycle. A cycle is one complete revolution taken by the rotating conductor. These two extinction points are changeover points from being positive in one direction to negative in the other direction. They also indicate the fastest rate of change from one peak to the other. Since *frequency* (*f*) is the number of cycles per second, the unit being the hertz (Hz), the time taken for one cycle to occur is called the *periodic time* (*T*). This can be expressed as:

$$T = 1/f \qquad (3.1)$$

It should be mentioned that the public electricity supply in the UK is generated at a frequency of 50 Hz, i.e. fifty cycles per second. So for one cycle to occur, the periodic time is 1/50 which is only 0.02 s and should give the reader an idea of how fast an a.c. generator must be driven to cram in fifty conductor revolutions in one second. This speed (*n*) can be found from the formula:

$$n = f/p \qquad (3.2)$$

where *p* is the generator's magnetic field pole pairs.

If the generator only had one set of north and south poles (one pole pair), in order for it to generate 50 Hz it would have to be driven at 50 rev/s or 3000 rev/min. Furthermore, at this frequency the supply is automatically being switched off 100 times every second. It will also be seen from the graph of Figure 1.10 that the cyclic nature of the a.c. sinewave, momentarily increases beyond the line marked 0.707. This is the value which refers to the *root mean square value* (r.m.s.) or *effective value* of the sinewave and is a measure of its usefulness. For comparison with direct current, it is the r.m.s. value which would produce the same heating effect. All a.c. supply voltages and their currents are measured in r.m.s. values, for example a domestic single-phase supply of 240 V or a three-phase supply of 415 V. It should be noted that for, say a 240 V a.c. supply the sinewave reaches a maximum value of 339 V. The method of determining the r.m.s. value has already been given in Chapter 1.

Circuit components

In a.c. theory we are interested in the behaviour and property of three circuit components, namely a resistor's *resistance*, an inductor's *inductance*, and a capacitor's *capacitance*. Electrical equipment may possess one or more of these properties such as a motor winding which has resistance and inductance or a discharge lamp circuit having an inductive ballast. It is the effect these components have on the relationship between supply current and supply voltage that causes problems in a.c. circuits. With the exception of a purely resistive circuit it is usually found that inductive and capacitive components create a *phase displacement* which occurs between current and voltage and this cannot be allowed to go unnoticed. Any marked phase shift will ultimately be detrimental to the supply system leading to higher costs. This will be explained later in the chapter.

Unlike a d.c. supply where power is the product of voltage and current (i.e. $P = VI$), a phase displacement in an a.c. supply puts the current out of step with the voltage. This phase displacement is represented by a *phase angle* and the cosine of this phase angle (cos ϕ) is called the *power factor* (see Figure 3.1). The power consumed in a single-phase a.c. circuit is not just $P = VI$ but is the product of three factors, namely, voltage, current and power factor (cos ϕ) and this can be expressed as:

$$P = V \times I \times \text{p.f.} \tag{3.3}$$

By transposition of the formula it can be seen that:

$$\text{p.f.} = P/VI$$

Power factor will be dealt with under the following sub-headings and later in the chapter but at this stage it is sufficient to say that it has no unit like the ampere or volt but it results in one of three conditions.

Resistance (R)

A resistor is a component having resistance as its chief property and as already mentioned its unit is the ohm (Ω). When a resistor is connected to an a.c. source as shown in Figure 3.2(a) it presents the only opposition to current flow. In this component the same rules used for Ohm's Law apply since the current and voltage are both effective values and their sinewaves pass through the same instantaneous points together. This waveform is shown in Figure 3.2(b) and from it can be constructed a phasor diagram to indicate the magnitude and direction of rotation of both these voltage and current quantities. It should be pointed out that rotation is usually in an anticlockwise direction and whilst the phasor diagram can be drawn at any angle within the confines of a circle with its fixed point being at the centre of the

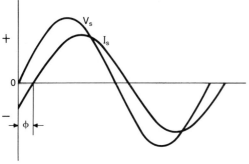

Fig. 3.1 Phase displacement between supply voltage and supply current.

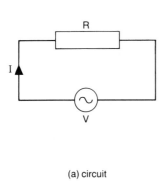

(a) circuit

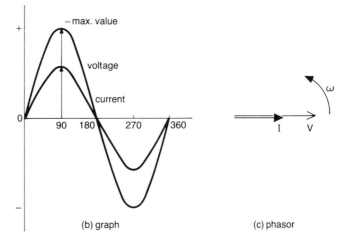

(b) graph

(c) phasor

Fig. 3.2 Circuit/graph/phasor for 'purely' resistive a.c. circuit.

circle, it is normally the practice to draw a phasor diagram in the horizontal position making one of the quantities a reference point.

The voltage and current quantities in Figure 3.2(c) are meant to be superimposed on each other but are drawn side by side for clarity. It will be seen that there is no phase angle between voltage and current and they are said to be *in-phase* with each other. Since there is no phase angle then one will find that cos 0° = 1 and since 1 is unity the power factor of a resistive component is said to be *unity power factor* (u.p.f.).

From the expression given earlier (3.3),

$$P = V \times I \times 1$$

which is the same as that for d.c. power. Interestingly, if the voltage and current instantaneous values are multiplied together, the resulting power curve would look like Figure 3.3(a). It can clearly be seen that a resistive component consumes power. Equipment possessing resistance such as filament lamps, electric fires, water heaters, etc. are able to utilize this power to create heat energy but in most other cases where resistance is found such as in long conductors and motor windings and coils, the consumed power is wasteful and represents a form of power loss in a circuit.

Inductance (L)

An *inductor* is a component having inductance as its chief property and as already mentioned its unit is the henry (H). Because an inductor is a coil it requires to be wound with resistance wire; inductance is not its only property. The opposition to current flow is both resistance and inductance and the combined effect is called *impedance* (Z). The unit of impedance is the ohm. However, in order to express the inductance property in ohms, the term inductive reactance (X_L) is used.

Inductive reactance is independent of resistance and is given by the formula:

$$X_L = 2\pi fL \tag{3.4}$$

In order to find an inductive circuit's impedance, one can either use an ammeter and voltmeter and use the formula $Z = V/I$ or use the formula:

$$Z = \sqrt{R^2 + X_L^2} \tag{3.5}$$

Note: This formula has been used in the previous chapters.

Figure 3.4 illustrates the condition in an a.c. circuit if the inductor had no resistance, often described as a 'pure' inductor. It will be seen that the current quantity *lags* behind the voltage quantity by a phase angle of 90°. This phase displacement of current is caused by the induced voltage created by the inductor opposing the supply voltage and was explained in the previous chapter. Under these conditions the circuit would have a *lagging power factor* and if the instantaneous voltage and current quantities were again multiplied together over one complete cycle it would be seen that no power is consumed (see Figure 3.3(b)). This is because the positive and negative quarter cycles of power cancel each other and also because cos 90° = 0. From the expression (3.3) it will be seen that $P = V \times I \times 0 = 0$ watts.

In practice, where coils are designed to be highly

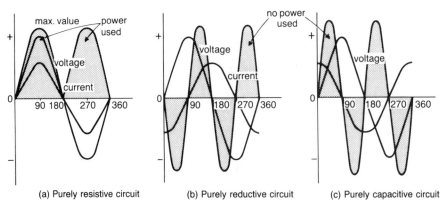

(a) Purely resistive circuit   (b) Purely reductive circuit   (c) Purely capacitive circuit

Fig. 3.3 Power curves.

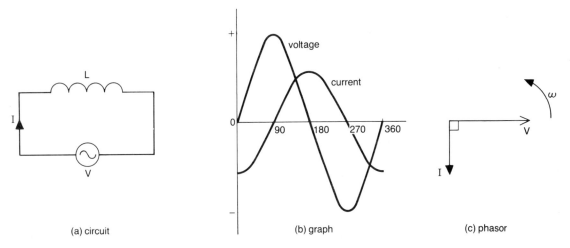

Fig. 3.4 Circuit/graph/phasor for 'purely' inductive a.c. circuit.

inductive, it means that they possess neglible resistance and they have extremely poor power factors. It should also be noted that such inductive coils consume very little power, their purpose in a.c. circuits is to produce artificial electromagnetism either to create magnetic flux or to create voltage changes to a higher or lower level than the normal a.c. supply voltage.

Capacitance (C)

A *capacitor* is a component having capacitance as its chief property and the unit of capacitance is the *farad* or *microfarad* (µF). When the capacitor is connected to an a.c. supply its plates are continually being charged and discharged owing to the positive and negative cycles of the a.c. supply. In actual fact, no current flows through the capacitor but as explained in the previous chapter, current flows first before the voltage on the plates is established. For a 'purely' capacitive circuit, the voltage and current waveforms are illustrated in Figure 3.5 where it will be seen that the current quantity *leads* the voltage quantity by 90°.

The capacitor is a component which creates a *leading power factor*. Its effect in a circuit is opposite to that of the 'pure' inductor and by superimposing the power quantity on to the voltage and current quantities it will be seen that the capacitor consumes no power. If a voltmeter and ammeter were connected in the capacitor's circuit and Ohm's Law

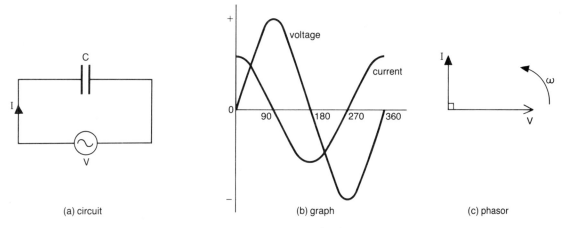

Fig. 3.5 Circuit/graph/phasor for capacitor connection in a.c. circuit.

applied, the ratio voltage/current would not be resistance but instead *capacitive reactance (X_C)* and like inductive reactance its unit is also the ohm. The term capacitive reactance can be found from the formula:

$$X_C = 1/(2\pi fC) \tag{3.6}$$

The expression (3.5) for finding impedance would be used only if the capacitor were connected with a resistor and/or inductor.

It is very important for students to remember the *three* basic phasor diagrams if they are to understand mixed circuit components. Students must also remember to treat *RLC* circuit graphical symbols in their 'pure' state unless told otherwise that they are a mixture of each or that they have definite angle of lag or lead.

Resistance and inductance in series

Example

Figure 3.6 shows a circuit diagram of a resistor of 30 Ω connected in series with an inductor of neglible resistance having an inductive reactance of 40 Ω. If the supply to the circuit is 250 V and the frequency 50 Hz, determine:

a) the impedance of the circuit
b) the current in the circuit
c) the p.d. across each component
d) the power factor of the circuit
e) the inductance of the coil

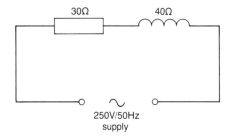

30Ω   40Ω

250V/50Hz
supply

Fig. 3.6 *RL* Series circuit.

Draw a phasor diagram of the circuit using the current as a reference line for drawing the voltages.

Solution

a) Use the impedance
    formula: $Z = \sqrt{R^2 + X_L^2}$
    then $Z = \sqrt{30^2 + 40^2} = \underline{50\ \Omega}$
b) Since $I = V/Z$
    then $I = 250/50 = \underline{5\ A}$
c) The p.d.s across the resistive and inductive parts of the circuit are respectively:
       $V = IR\ \ = 5 \times 30 = \underline{150\ V}$
    and $V = IX_L = 5 \times 40 = \underline{200\ V}$
    It should be noted that the algebraic sum of the p.d.s is 150+200 = 350 V yet the supply is 250 V. Reference to the phasor diagram will explain that in a.c. mixed circuits, the supply voltage (or current) is the *phasor sum* and not the algebraic sum.
d) The power factor of the circuit will be lagging since the coil is inductive and this can be found by the cosine ratio between resistance and impedance, i.e. cos φ = R/Z. It could also be found from the voltages in the phasor diagram, i.e. cos φ = V_R/V_S. Using the first method, the the power factor cos φ = R/Z = 30/50
    = $\underline{0.6\ lagging.}$
e) The inductance of the coil is found by using formula (3.4) and rearranging it to find *L*.

       Since $X_L = 2\pi fL$
       then $L = X_L/2\pi f$
                   = 40/314.2
                   = $\underline{0.127\ H}$

The phasor diagram is shown in Figure 3.7. It should be noted that the current is common to both circuit components and this is the reason why it was chosen as the reference line. Also note that the current is in phase with the resistive component's own p.d. and that the p.d. across the inductive component is 90° out-of-phase with the reference current. By drawing a parallelogram of these two p.d.s the diagonal line will represent the supply voltage. Students should fully label their diagram, inserting in it the phase angle and the direction of rotation.

Figure 3.8 shows how the circuit components can be drawn to represent an impedance diagram since their common unit is the ohm. Both this diagram and the voltage part of the phasor diagram are derived at by applying Pythagoras's theorem since a phase angle of 90° has been assumed for the inductor.

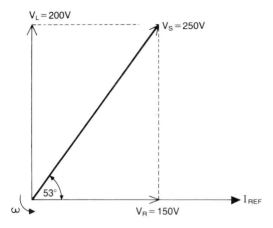

Fig. 3.7 Phasor diagram.

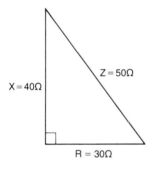

Fig. 3.8 Impedance triangle.

The above example shows that the circuit is not highly inductive but a power factor of 0.6 lagging is poor. It will be explained later that a power factor approaching 0.85 lagging is more realistic.

In practice, an inductive coil is a mixture of resistance and inductance and it would be impossible to separate the internal p.d.s. On an a.c. supply a voltmeter connected across the coil and an ammeter in series with it would yield only the coil's impedance. It is possible, however, to connect the coil to a d.c. supply and determine its resistance only. The inductive reactance and hence inductance can then be found.

Example

Figure 3.9 shows an inductive coil connected to an a.c. supply of 240 V, 50 Hz. An ammeter in the circuit reads 2 A. If the coil is connected to a d.c. supply of 100 V the ammeter reads 10 A. What is the coil's
(a) inductive reactance
(b) inductance?

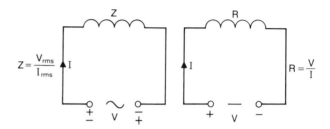

Fig. 3.9 Coil connected to a.c. and d.c. supply.

Solution

On a.c. $\qquad Z = V/I = 240/2 = \underline{120\ \Omega}$

On d.c. $\qquad R = V/I = 100/10 = \underline{10\ \Omega}$

Re-arranging (3.5) $X_L = \sqrt{(Z^2 - R^2)}$
$$= \sqrt{(120^2 - 10^2)}$$
$$= \underline{119.6\ \Omega}$$

Re-arranging (3.4) $\ L = X/2\pi f$
$$= 119.6/314.2$$
$$= \underline{0.38\ H}$$

Note that this coil is highly inductive and its power factor on the a.c. supply is found by:

p.f. $= R/Z$
$$= 10/120 = \underline{0.08\ lagging}$$

This is a very poor power factor, causing the a.c. current to lag behind the supply voltage by a phase angle of 85°. See the phasor diagram in Figure 3.10.

Resistance and capacitance in series

Example

Figure 3.11 shows a 50 μF capacitor connected in series with a 60 Ω non-inductive resistor across a 240 V, 50 Hz supply. Determine:
a) the capacitive reactance of the capacitor

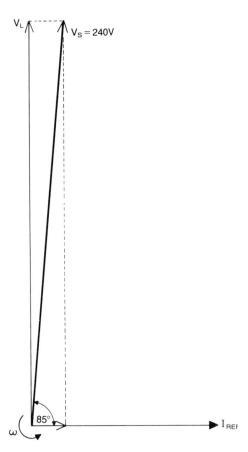

Fig. 3.10   Phasor diagram for highly inductive circuit.

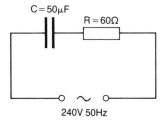

Fig. 3.11   *RC* Series circuit.

b) the impedance of the circuit
c) the current taken by the circuit
d) the power factor and phase angle of the circuit
e) the power consumed.
Draw a phasor diagram showing the current and voltage relationship.

Solution

a)   From (3.6) $X_C = 1/(2\pi f C)$
       Since C is in μF then $X_C = 10^6/(314.2 \times 50)$
                                              $= 63.65\ \Omega$

b)   The impedance is $Z = \sqrt{(R^2 + X_c^2)}$
                                              $= \sqrt{(60^2 + 63.65^2)}$
                                              $= \underline{87.47\ \Omega}$

c)   The circuit current is $I = V/Z$
                                              $= 240/87.47$
                                              $= \underline{2.74\ A}$

d)   The power factor is $\cos\phi = R/Z$
                                              $= 60/87.47$
                                              $= \underline{0.68\ leading}$
       From this the phase angle $\phi = \underline{46.69°}$

e)   The power consumed is given by formula (3.3)
       Thus $P = VI\cos\phi$
                                              $= 240 \times 2.7 \times 0.68$
                                              $= \underline{447.2\ W}$

The phasor diagram is shown in Figure 3.12. It should be noted that the resistor in circuit reduces the phase angle between the supply current and supply voltage, i.e. if it wasn't connected the current would lead the voltage by 90°.

Resistance and inductance in parallel

Example

Figure 3.13 shows a non-inductive resistor of 40 Ω connected in parallel with a coil of inductance 95.6 mH and negligible resistance. If the supply voltage is 240 V, 50 Hz, determine:

a)   the inductive reactance of the coil
b)   the current through each circuit component
c)   the supply current, power factor and phase angle.
       Draw a phasor diagram of the circuit using a suitable scale.

Solution

a) $X_L = 2\pi f L$
              $= \underline{314.2 \times 0.0956}$
              $= \underline{30\ \Omega}$

b) $I_R = 240/40 = \underline{6\ A}$
       $I_L = 240/30 = \underline{8\ A}$

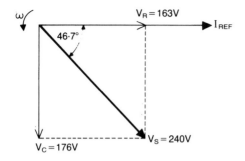

Fig. 3.12 Phasor diagram for *RC* Series circuit.

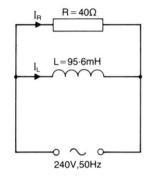

Fig. 3.13 *RL* parallel circuit.

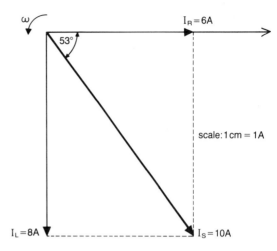

Fig. 3.14 Phasor diagram for *RL* parallel circuit.

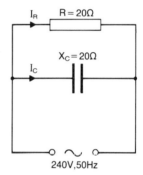

Fig. 3.15 *RC* parallel circuit.

c) See Figure 3.14.

The supply current $I_s = \sqrt{(I_R^2 + I_L^2)}$
$$= \sqrt{6^2 + 8^2} = \underline{10 \text{ A}}$$

Since power factor $\cos \phi = I_R/I_s$
$$= 6/10 = \underline{0.6 \text{ lagging}}$$

then phase angle $\phi = \underline{53°}$

Resistance and capacitance in parallel

Example

Figure 3.15 shows a non-inductive resistor of 20 Ω connected in parallel with a capacitor having a capacitive reactance of 20 Ω. If the supply voltage is 240 V, 50 Hz determine:

a) the capacitor's capacitance
b) the current through each component
c) the supply current, power factor and phase angle.

Draw a phasor diagram of the circuit using a suitable scale.

Solution

a) $C = 1/(2\pi fX)$
$$= 10^6/(314.2 \times 20)$$
$$= \underline{159 \text{ μF}}$$

b) $I_R = 240/20 = \underline{12 \text{ A}}$
$I_C = 240/20 = \underline{12 \text{ A}}$

c) $I_S = \sqrt{(I_R^2 + I_C^2)}$
$$\sqrt{(12^2 + 12^2)}$$
$$= \underline{16.97 \text{ A}}$$

$\cos \phi = I_R/I_S$
$$= 12/16.97 = \underline{0.707 \text{ leading}}$$

$\phi = \underline{45°}$

See Figure 3.16

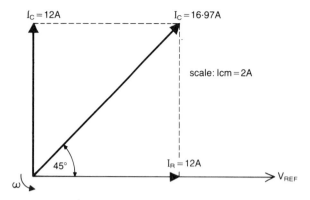

Fig. 3.16 Phasor diagram for *RC* parallel circuit.

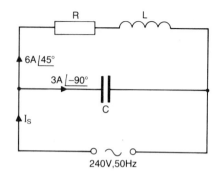

In practice, it is often an inductor in an a.c. circuit that requires the use of a capacitor to correct the phase angle. The following example will illustrate this.

Example

Figure 3.17 shows an inductive coil in parallel with a capacitor. If the inductor takes a current of 6 A and this current lags behind the supply voltage by a phase angle of 45°, what will be the current taken from the supply if the capacitor's current is 3 A and leads the supply voltage by 90°?

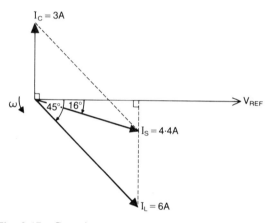

Fig. 3.17 Capacitor.

Solution

Draw a horizontal line to indicate the reference voltage (this is chosen because in parallel circuits the voltage is common across the branch components). Choose a suitable scale (1 cm = 1 A) and with ruler and protractor draw the inductor's current of 6 A lagging the supply voltage by 45°. Now draw the capacitor's current of 3 A leading the supply voltage by 90°. Construct a parallelogram and insert the diagonal line. This line represents the supply current and is measured to be 4.4 A lagging behind the supply voltage by 16°. The phasor diagram indicates that the power factor of the circuit has been improved from 0.707 lagging to 0.96 lagging. It also shows that instead of taking 6 A from the supply, the coil with a parallel connected capacitor now only takes 4.4 A.

It should become a little clearer to students that inductors and capacitors have the effect of neutralizing each other when they are connected to a.c. supplies. Let us consider the following example of a circuit containing all three components in the 'pure' state.

Example

A series circuit comprises a resistor of 5 Ω, inductor 0.02 H and capacitor 150 μF. If components are connected to a single-phase a.c. supply of 240 V, 50 Hz, what are the supply current and power factor?

Solution

The circuit diagram is shown in Figure 3.18. The procedure is to firstly find the inductor's and capacitor's reactance values expressed in ohms.

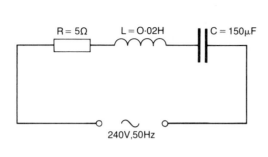

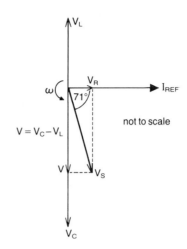

Fig. 3.18 *RLC* Series circuit.

Thus from (3.4), $X_L = (2\pi fL) = 314.2 \times 0.02 = \underline{6.28\ \Omega}$
Also from (3.6) $\quad X_C = 1/(2\pi fC) = 10^6/(314.2 \times 150)$
$$= \underline{21.2\ \Omega}$$
These reactances are in anti-phase with each other
(i.e. they are 180° apart and in opposite directions)
and they must be subtracted.
The resultant reactance is leading
$$X = X_C - X_L$$
$$= 21.2 - 6.28 = \underline{14.92\Omega}$$
The impedance of the circuit is given by
$$Z = \surd(R^2 + X_C{}^2)$$
$$= \surd(5^2 + 14.92^2) = \underline{15.75\ \Omega}$$
The supply current $I = V/Z$
$$= 240/15.75 = \underline{15.2\ A}$$
The power
factor $\quad\quad \cos\phi = R/Z = 5/15.75 = \underline{0.32\ \text{leading}}$

Power factor

In general, power factor can be one of three
conditions, namely:

1   unity power factor
2   lagging power factor
3   leading power factor

Examples of these conditions have already been given
and resulted from the type(s) of component found in
a.c. circuits, whether they were resistive, inductive or
capacitive. A purely resistive load would result in the
unity of p.f. condition with the above ratio $P/VI = 1$.
The other two conditions are created by inductive

and capacitive components and will make power
factor less than 1. This infers that power factor
cannot exceed 1 since cos 90° = 0 and no power will
be taken from the supply.

In the examples previously given the phasor
diagrams have been constructed from the knowledge
of an in-phase quantity and a quadrature phase or
reactive quantity – these quantities being either
current or voltage. The reactive quantity is at 90° and
refers to the component which consumes no power.
For this reason it is usually the practice to express
power factor as a ratio between the *true power* in the
circuit (watts) and the *apparent power* (voltamperes),
that is:

> Power factor (cos ϕ) = true power (P)/apparent
> power (S)

A wattmeter measures the true power and a
voltmeter and ammeter arrangement measures the
apparent power. The line representing the no-power
quantity of the inductor or capacitor is called the
reactive voltamperes (Q). Figure 3.19 shows the
power triangles commonly used:

Example

Figure 3.20 shows the circuit connections of a single-
phase a.c. motor. The wattmeter reads 5 kW, the
voltmeter reads 240 V and the ammeter reads 32 A.
Draw the power triangle and determine the power
factor of the circuit and the reactive voltamperes.

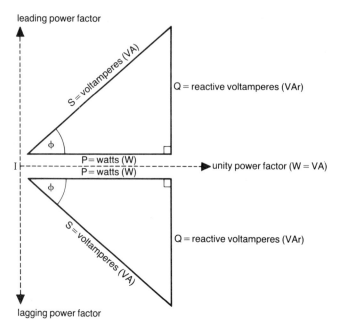

Fig. 3.19 Power triangles showing power factor conditions.

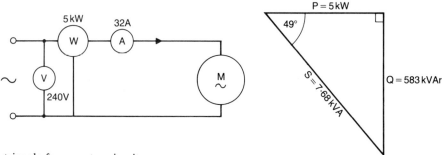

Fig. 3.20 Power triangle for a motor circuit.

Solution

See Figure 3.20 showing the power triangle, it will have a lagging p.f.

Power factor

$(\cos \phi) = P/S = 5000/(240 \times 32) = \underline{0.65 \text{ lagging}}$

Since $S = \sqrt{(P^2 + Q^2)}$

then $Q = \sqrt{(S^2 - P^2)}$

                $= \sqrt{(7.68^2 - 5^2)} = \underline{5.83 \text{ kVAr}}$

Three-phase supplies

At this stage the emphasis has been on *RLC* components, in order to explain their behaviour in relationship with the supply voltage and circuit current. Power factor was introduced as a result of phase displacement between voltage and current waveforms. Attention must now be paid to the public system which for most users is obtained from a *three-phase, four-wire* system.

Generation of three-phase alternating current is created by the principles of electromagnetic induction as was the single-loop conductor given earlier. Figure 3.21 shows how the three loops are situated and equally spaced at 120° intervals. For clarity the magnetic field has been omitted. These conductors are identified as the *red* phase, *yellow* phase and *blue* phase respectively and it will be seen that when one of these phases reaches maximum in any direction the other two phases become not only opposite in polarity but also half the magnitude.

It is normal practice in the power station to *star-connect* one end of the three-phase conductor loops and earth this point to create stability within the system. At some point in the supply distribution network the three-phase will obtain a *neutral conductor*. The reason for doing this is because there is no guarantee that the system will be *balanced*. Any of the phases may take a load current different from

the other two phases and by introducing a neutral conductor, out-of-balance loading will return current along this path.

It will be seen in the diagram that the system is in fact balanced and no current flows through the neutral. At the instantaneous positions shown, the yellow-phase conductor is sending current (say 30 A) and the red and blue phases are returning half this value of current (15 A). It should be pointed out that the instantaneous values chosen are not r.m.s. values; an ammeter connected in the circuit would indicate the r.m.s. value and all the phases would show the same reading (i.e. if the system was balanced).

When dealing with three-phase electrical apparatus, one will come across connections known as *star* and *delta* and the term *root three* ($\sqrt{3}$) will be commonly used. This term distinguishes between *line* and *phase* values as a result of the star and/or delta connection. To illustrate this, Figure 3.22(a) shows the phasor

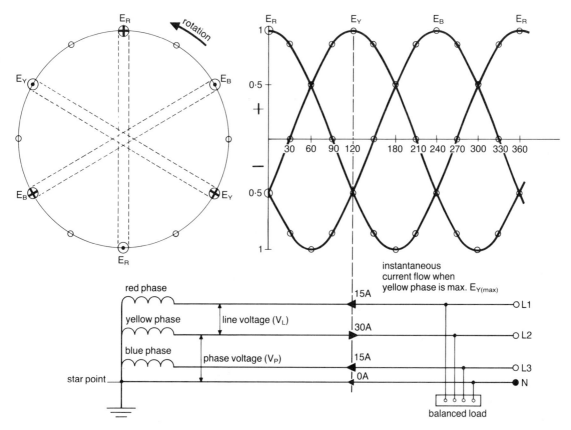

Fig. 3.21 Three-phase generation of electricity.

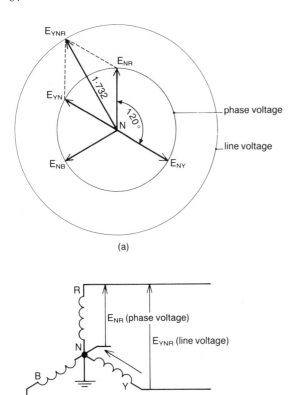

(a)

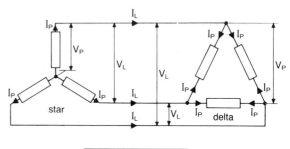

(b)

Fig. 3.22 Line voltage and phase voltage in a star-connected supply system (a) phasor diagram (b) star connection.

diagram associated with a star connected system. Only the red phase voltage and red-to-yellow line voltage will be dealt with for clarity.

Firstly, consider the phasor diagram showing the induced voltages for each of the phases. These are lettered accordingly from the neutral point outwards, for example, neutral-to-red phase voltage is called E_{NR}. This voltage is shown in Figure 3.22(b) and is acting across the red phase. If, however, a voltmeter were placed across the red line and yellow line it would not read the same value as the phase voltage. To understand this it will be seen that the yellow phase-to-neutral voltage E_{YN} has to be considered and added to E_{NR} vectorily not algebraically. This is shown in Figures 3.22(a) to be equal and opposite in the direction it normally acts.

From these two phase voltages is constructed a parallelogram which produces the actual line value between the red and yellow phases, i.e. E_{YNR}. The length of this line is found to be 1.732 ($\sqrt{3}$) longer than the phase value. Students ought to complete Figure 3.22(a) for the other two line voltages. The relationship between the ($\sqrt{3}$) value for a star and delta system are shown in Figure 3.23 and it is important for students to remember the difference between line and phase values for the two forms of connection. The following example illustrates this difference.

	star	delta
	$I_L = I_P$	$I_L = \sqrt{3}\,I_P$
	$V_L = \sqrt{3}\,V_P$	$V_L = V_P$

Legend:
V_L is line voltage
V_P is phase voltage
I_L is phase current
$\sqrt{3}$ is 1·732

Fig. 3.23 Star and delta connections.

Example

Figure 3.23 represents a three-phase, star-connected supply system feeding a delta-connected load. If the star-connected phase voltage is 240 V and the phase current is 20 A, determine the line values of voltages and current and also the delta line and phase values of voltage and current.

Solution

In the star-connected system

$I_L = I_P,$ therefore $I_L = 20$ A
$V_L = \sqrt{3}V_P,$ therefore $V_L = \sqrt{3} \times 240 = 415$ V

In the delta-connected system

$I_L = \sqrt{3}I_P,$ therefore $I_P = 20/\sqrt{3} = 11.56$ A
$V_L = V_P,$ therefore $V_L = 415$ V

Note: In practice students will come across a delta–star connected transformer where the supply is reduced from 11 kV to 415/240 V. This will be discussed in Chapter 5 dealing with transformers.

The power in a three-phase system is given by the expression:

$$P = \sqrt{3}\, V_L I_L \cos \phi \qquad (3.7)$$

The following examples illustrate how three-phase power can be determined by calculation.

Example 1

Calculate the line current and total power consumed by three 40 Ω resistors connected in star to a 415 V three-phase star supply system.

Solution

The phase voltage across the load is
$$V_P = V_L/\sqrt{3}$$
$$= 415/1.732 = \underline{240\ V}$$
The line current is the same as the phase current, i.e.
$$I_P = I_L$$

But $I_P = V_P/R$
$$= 240/40 = \underline{6\ A}$$
Since the resistors are equal, this will be the current in the other lines. Also note that resistors produce unity power factor.
From the formula (3.7) $P = \sqrt{3}\, V_L I_L \cos \phi$
$$= 1.732 \times 415 \times 6 \times 1$$
$$= \underline{4313\ W}$$

Example 2

Three impedances each having a value of 40 Ω and power factor 0.8 lagging are to be connected in (a) star and (b) delta across a 415 V three-phase supply. Determine the supply line current and the total power consumed.

Solution

a) In star $I_P = 240/40 = \underline{6\ A}$
$$P = \sqrt{3} \times 415 \times 6 \times 0.8 = \underline{3450\ W}$$
b) In delta $I_P = 415/40 = \underline{10.375\ A}$
  Since $I_L = \sqrt{3}I_P = \sqrt{3} \times 10.375 = \underline{18\ A}$
$$P = \sqrt{3} \times 415 \times 18 \times 0.8 = \underline{10\ 350\ W}$$

The above example shows that the power consumed in delta is three times as much as in star. Since the power is transformed into heat energy, it is for this reason that a three-phase induction motor is sometimes started in star–delta. It is to reduce the supply voltage at starting to 240 V (star connection) before switching the windings for the full 415 V (delta connection). The heat dissipated in the windings by starting this way is greatly reduced. Electrical machines will be dealt with in Chapter 4.

Example 3

In a 415 V three-phase, four-wire system (see Figure 3.24), three consumers have the following three-phase loads:
(a) Consumer A takes 50 kW at unity power factor.
(b) Consumer B takes 80 kVA at 0.6 power factor lagging.
(c) Consumer C takes 40 kVA at 0.7 leading power factor.
Determine the systems overall kW, kVA, kVAr and power factor.

Solution

(a) Consumer A:
  Since $\cos \phi = P/S$
$$P = \underline{50\ kW}$$
$$S = P/\cos \phi = 50/1 = \underline{50\ kVA}$$
$$Q = 0$$
(b) Consumer B:
$$P = S \times \cos \phi = 80 \times 0.6 = \underline{48\ kW}$$
$$S = 80\ kVA$$
$$Q_B = S \times \sin \phi = 80 \times 0.8 = \underline{64\ kVAr}$$
(c) Consumer C:
$$S = 40\ kVA$$
$$P = S \times \cos \phi = 80 \times 0.6 = \underline{48\ kW}$$
$$Q_C = S \times \sin \phi = 40 \times 0.714 = 28.57$$
$$\underline{kVAr}$$

The total kilowatts used by the three consumers is:
$$P = 50 + 48 + 28 = \underline{126\ kW}$$

The total kVAr is the difference between the leading and lagging power factor conditions.

Thus $Q = Q_B - Q_C = 64 - 28.57$
$$= \underline{35.43\ kVAr}$$

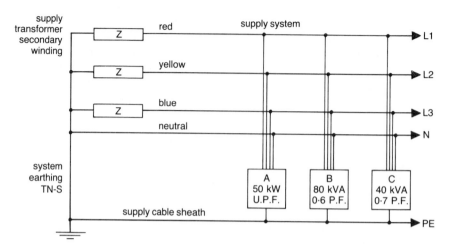

Fig. 3.24 Three-phase four wire supply.

The total kVA can be found by Pythagoras's theorem:

$$S = \sqrt{(P^2 + Q^2)}$$
$$= \sqrt{(126^2 + 35.43^2)}$$
$$= 130.9 \text{ kVA}$$

The overall power factor is $\cos \phi = P/S$
= 126/130.9
= 0.96 lagging

The phasor diagram of the system is shown in Figure 3.25.

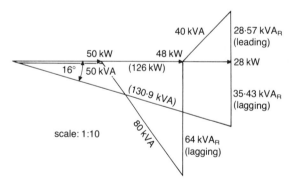

Fig. 3.25 Graphical solution.

Exercise 3

1   (a)   For the circuit shown in Figure 3.26 the value of R is 12 Ω and the value of X_L is 16 Ω. Calculate:

       (i)   the impedance of the circuit

      (ii)   the current flowing

     (iii)   the potential difference across each component

  (b)   Draw the phasor diagram of the circuit and determine the phase angle and power factor.

CGLI/II/83 (Mod)

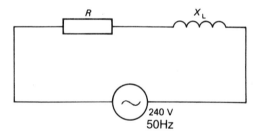

Fig. 3.26 Inductive circuit.

2   A factory supply is 11 kV, three-phase, 3-wire and is fed to a transformer whose output is 415 V/240 V three-phase, 4-wire.

  (a)   Draw the circuit diagram of this system.

  (b)   What is the purpose of earthing the star point of the transformer?

  (c)   Determine the total kVA load on the transformer if the following three phase loads were connected.

      (i)   120 kW heating load at unity power factor

(ii) 240 kVA load at 0.8 power factor lagging

CGLI/II/80 (Mod)

3 If the capacitor in Figure 3.27 had a value of 100 μF and the supply was 240 V, determine:
   (a) the circuit current if R and X_L were 8 Ω and 15 Ω respectively
   (b) the power and power factor of the circuit.

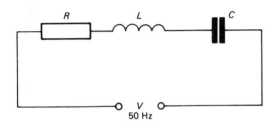

Fig. 3.27 *RLC* series circuit.

4 Draw a phasor diagram of Figure 3.28 and find the value of the supply current and power factor.

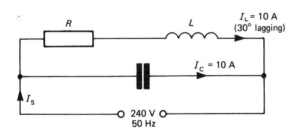

Fig. 3.28 Mixed circuit components.

5 A 50 kW a.c. motor operates at a power factor of 0.65 lagging and efficiency 83%. A 40 kVAr power factor improvement capacitor is to be connected in parallel with the motor. Find graphically or by calculation:
   (a) the full load kVA before the improvement capacitor is connected
   (b) the supply kVA after the improvement capacitor is connected.

CGLI/II/82 (Mod)

6 Figure 3.29 shows the connections of a SON discharge lamp. With the switch S open, ammeter A_1 reads 5 A and the wattmeter W reads 420 W. With S closed A_2 reads 2.2 A and A_1 reads 3 A. the wattmeter reading is unaltered.
Draw a phasor diagram of the circuit using a scale 1 A = 2 cm.

7 Three inductive coils of resistance 12 Ω and reactance 9 Ω are connected in star to a 440 V, 50 Hz supply. Determine:
   (a) the line current
   (b) the power factor
   (c) the total power
   (d) the apparent power

8 Make a sketch of sinewave voltage and current over one complete cycle. The sinewave voltage reaches a maximum value of 50 V while the current sinewave reaches a maximum value of 10 A and it lags behind the sinewave voltage by an angle of 60°.

9 In question 8 above, determine from the graph the r.m.s. value of the voltage waveform using the mid-ordinate rule.

10 (a) Explain why it is important to balance a.c. loads.
   (b) What is the benefit of providing a three-phase supply system with a neutral return conductor?
   (c) State the polarity of a three-phase, four-wire supply system including earthing arrangements.

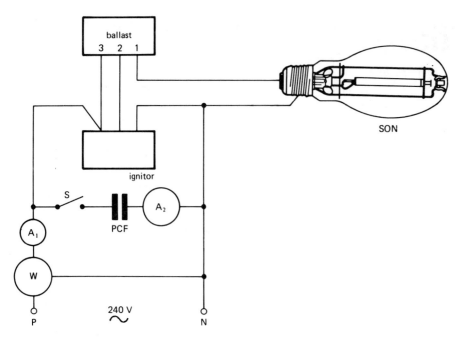

Fig. 3.29 SON discharge lamp circuit.

CHAPTER FOUR

Electric motors

After reading this chapter you will be able to:

1 State briefly the operation of these motors:
   - (*a*) three-phase cage induction motor
   - (*b*) single-phase, split-phase motors
   - (*c*) single-phase shaded pole motor
   - (*d*) universal motor
   - (*e*) repulsion motor
   - (*f*) d.c. series, shunt and compound motors

2 Explain the essential difference between motors that operate on *induction* and those that operate on *conduction*.

3 Perform calculations associated with the operation and performance of the motors listed above.

4 Draw circuit diagrams of the motors listed above.

An introduction to motor installations is given in the book *Theory and Regulations* by the same author and it is the intention of this chapter to deal with the operation of different types for use on a.c. and d.c. supplies.

If the principles of electromagnetic induction are fully understood by students, together with the knowledge that force is exerted on a conductor carrying current situated in magnetic field, then an understanding of most motors will not present too much difficulty. The book mentioned above made a distinction between motors that run on a.c. supplies and those that run on d.c. supplies and this was simply the way two magnetic fields were allowed to be created. It is magnetic field interaction which is at the heart of all motor operation.

Three-phase induction motors

Since this type of motor is one of the most widely used of all a.c. motors, its operation and general performance characteristics will be discussed first. A typical part view of this motor is shown in Figure 4.1 and the items marked 9 and 11 are the two essential features of this and any motor, namely the *rotor* and *yoke* (or *frame*). Whilst not indicated on the diagram, the yoke is made out of welded or rolled steel, and supports an inner *stator* which incorporates three main phase windings. These stator windings are insulated and placed symmetrically inside slotted laminations made from a high-grade alloy steel which are called *stampings*. Figure 4.2 shows two methods of insulating motor stator windings. In practice, motor windings are arranged to give either a star or delta connection and often item 24, the terminal box, will show six terminals.

The revolving part of the motor is called a *cage rotor* and as can be seen it is a cylindrical solid lump of metal. Embedded within this mass of metal will be found copper or aluminium bars which are joined at either end of the rotor by *end rings*. These rings effectively short-circuit the rotor bars. Like the stator, the rotor is also highly laminated to reduce eddy current losses (see items 30 and 31). These losses and other losses all reduce the motor's overall efficiency.

Two other types of rotor usually found are the (a) double cage rotor and (b) wound rotor. They are used to provide the motor with an increased starting torque. Figure 4.3 shows part of a double cage rotor. The outer cage has more resistance in it than the inner cage.

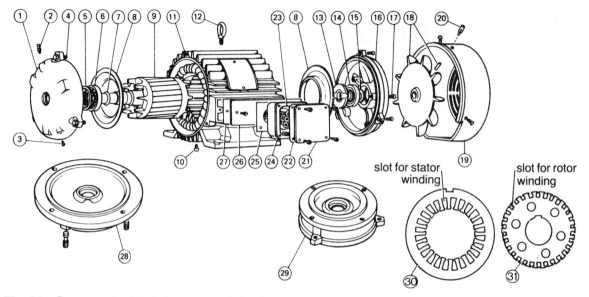

Fig. 4.1 Components of typical cage-rotor induction motor.

1	*Endshield, driving end*	*11*	*Yoke with or without feet*
2	*Grease nipple*	*12*	*Eyebolt*
3	*Grease relief screw*	*13*	*Inside cap, non-driving end*
4	*End securing bolt, or through bolt and nuts*	*14*	*Ball bearing, non-driving end*
		15	*Circlip*
5	*Anti-bump washers*	*16*	*Endshield, non-driving end*
6	*Ball bearing, driving end*	*17*	*Inside cap screws*
7	*False bearing shoulder*	*18*	*Fan with peg or key*
8	*Flume*	*19*	*Fan cover*
9	*Rotor on shaft*	*20*	*Lubricator extension pipe*
10	*Drain plug*	*21*	*Terminal box cover*

22	*Terminal box cover gasket*
23	*Terminal board*
24	*Terminal box*
25	*Terminal box gasket*
26	*Raceway plate*
27	*Raceway plate gasket*
28	*D flange*
29	*C face flange*
30	*Stator lamination*
31	*Rotor lamination*

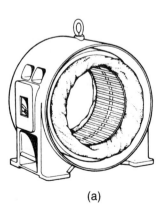

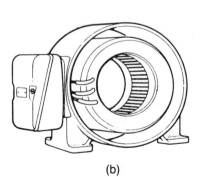

Fig. 4.2 Winding insulation (a) Oil modified polyester compound (b) Epoxy resin encapsulated.

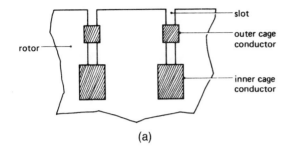

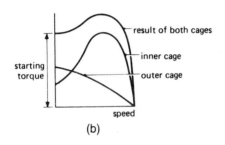

Fig. 4.3 (a) Sections of a double cage rotor
(b) Speed-torque characteristic of both cages.

Operation

The machine's three-phase windings – when connected to the a.c. supply – set up a rotating magnetic field around the stator core. To understand how this travelling field is established, reference should be made to Figure 4.4. The arrangement of the three-phase stator windings is shown in the top left circle.

Each phase is shown with a dot and a cross to signify the direction of the current (a dot indicates current coming out and a cross indicates current going in). The three other circles show what happens when each winding produces maximum current 120 electrical degrees later than the previous winding. The magnetomotive force (m.m.f.) direction of each circle can be brought out and shown as a flux phasor (see the flux phasor diagram). In order to understand the graph of the rotating magnetic field, it should be appreciated that the flux phasors are shown in the positive direction. This is important because the arrows must be reversed when each phase passes

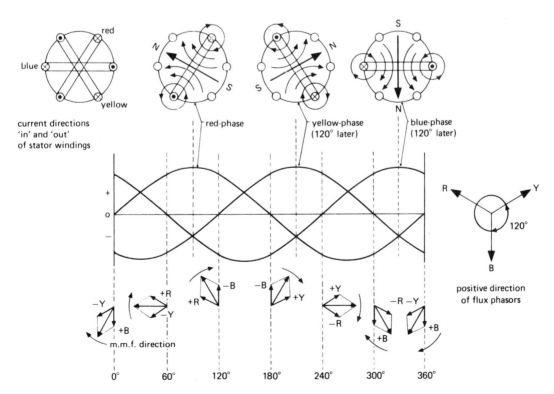

Fig. 4.4 Rotating magnetic field produced by the three-phase supply.

through zero and becomes negative. So that the field can be shown to move in a clockwise direction and to eliminate the problem of dealing with three phasors at a time, intervals of 60° have been chosen.

The rotating magnetic field travels around the stator core at *synchronous speed* (n_s). This speed is determined by the frequency of supply and number of pole pairs in the stator. The expression for finding it is:

$$n_s = \frac{f}{p} \tag{4.1}$$

where n_s is the synchronous speed in revolutions per second

f is the frequency of supply in hertz

p is the number of pole pairs

This travelling flux cuts through the rotor bar conductors causing e.m.f.s to be induced. Because the rotor bars are short-circuited by end rings, the e.m.f.s give rise to induced currents which in turn create their own magnetic fields. It is the interaction of these fields with the already established rotating magnetic field which is responsible for the rotor moving – the turning effort we call *torque*.

The interaction of both these fields is shown in Figure 4.5(a). The red phase m.m.f. is seen rotating in a clockwise direction, creating induced currents in the short-circuited rotor bars (the direction of these currents is found by applying *Fleming's right-hand rule*, Figure 2.16). The induced currents set up a rotor magnetic flux acting in the direction shown. From these two field interactions it will be seen that the rotor's north pole will repel the rotating field's north pole, or put another way, will be attracted towards the rotating magnetic field's south pole. The rotor will run in the same direction as the rotating magnetic field.

What is not often appreciated is that the rotor can never turn as fast as the travelling stator core flux. This is because the faster the rotor rotates, the less induced current there will be in the rotor bars. This means less interacting flux and less torque. If the rotor were travelling at the same speed as the rotating magnetic field, there would be no rotor flux, since no induced currents are created. It is, however, possible to have a rotor travelling at synchronous

Stage 1

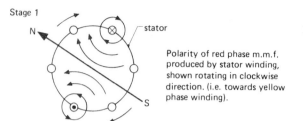

Polarity of red phase m.m.f. produced by stator winding, shown rotating in clockwise direction. (i.e. towards yellow phase winding).

Stage 2

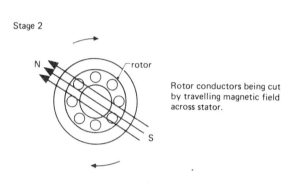

Rotor conductors being cut by travelling magnetic field across stator.

Stage 3

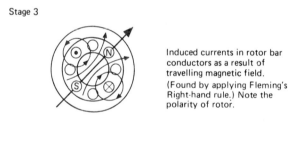

Induced currents in rotor bar conductors as a result of travelling magnetic field. (Found by applying Fleming's Right-hand rule.) Note the polarity of rotor.

Stage 4

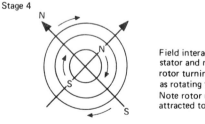

Field interaction between stator and rotor, showing rotor turning in same direction as rotating field. Note rotor north being attracted to stator south.

Fig. 4.5 (a) Induction motor principle.

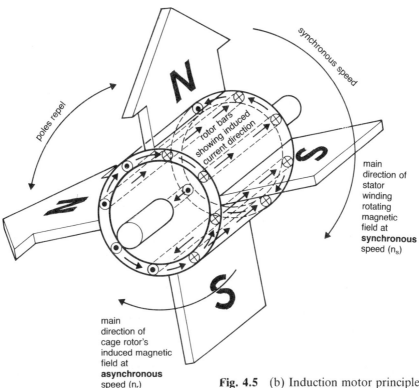

poles repel

N

rotor bars
showing induced
current direction

synchronous speed

main
direction of
stator
winding
rotating
magnetic
field at
synchronous
speed (n_s)

main
direction of
cage rotor's
induced magnetic
field at
asynchronous
speed (n_r)

Fig. 4.5 (b) Induction motor principles showing how torque is produced in a cage rotor.

speed by making it into a permanent magnet where its field can be 'locked' in with the travelling stator core field; a motor designed for this operation is called a *synchronous motor* – the induction motor is called an *asynchronous motor*.

Generally speaking, the rotor of the induction motor will settle down to a steady speed just below synchronous speed. It can therefore be regarded as a constant speed machine with an operating slip lying between 3% and 5%, i.e. 0.03 and 0.05 per unit. The term slip (s) is the difference between synchronous speed (n_s) and rotor speed (n_r). The ratio slip speed and synchronous speed is known a the per unit slip (s). i.e.

$$s = \frac{n_s - n_r}{n_s} \qquad (4.2)$$

Example 1

The synchronous speed of an induction motor is

25 rev/s and its rotor speed is 24.16 rev/s. Calculate the per unit slip.

$$\text{Since } s = \frac{slip\ speed}{synchronous\ speed}$$

$$\text{i.e. } s = \frac{n_s - n_r}{n_s}$$

$$s = \frac{25 - 24.16}{25}$$

$$= 0.033 \text{ p.u.}$$

Example 2

Determine the speed of a 6-pole induction motor which has a slip of 0.05 p.u. at full load. If the supply frequency is 50 Hz, what would be the speed of a 2-pole alternator (synchronous machine) supplying the motor?

The synchronous speed of the motor is

$$n_s = \frac{f}{p}(\text{where } p = 6/2 = 3)$$

$$= \frac{50}{3}$$

$$= 16.67 \text{ rev/s}$$

By transposing expression (4.2) the rotor speed is

$$n_r = n_s (1 - s)$$
$$= n_s - n_s s$$
$$= 16.67 - (0.05 \times 16.67)$$
$$= 15.84 \text{ rev/s}$$

The speed of the alternator is $f/p = 50/1 = 50$ rev/s

When the induction motor becomes mechanically loaded its speed (rotor speed) begins to fall and its slip increases. The frequency and value of rotor current increase and, as a result, more current is taken by the stator windings. It should be pointed out that when the rotor is stationary (at standstill), slip = 1. This implies that n_r (the rotor speed) is zero. Figure 4.6 is a typical torque-speed characteristic. The shaded portion represents the normal operating region and the 'knee' of the curve indicates the point where the motor *pulls-out* owing

to the torque increasing above a certain value (approximately 2.5 times full-load torque). The graph is not to scale. The dotted line represents the characteristic of an improved rotor cage to increase the starting torque – achieved with conductor bars having higher resistance. External rotor resistance and double-cage rotors are designed to improve the efficiency of this type of motor.

Single-phase induction motors

(a) Split-phase motors

The basic construction of these motors is similar to the three-phase, cage-rotor types. Instead of having a three-phase supply, which automatically provides the stator windings with a rotating magnetic field, split-phase motors operate on single-phase supplies. At first this would seem to present a problem if the motors had only one winding. This is because their stator windings will produce only a pulsating field, incapable of making their rotors turn. To overcome the problem, motors are designed with two separate windings, initially connected in parallel and arranged spatially to be 90° apart. One winding is called the

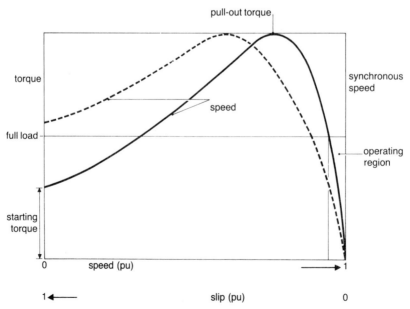

Fig. 4.6 Torque-speed characteristic of cage induction motor.

running winding (the main winding), the other called the *starting winding* (the auxiliary winding). The running winding is always connected in circuit but the starting winding may be disconnected.

To enable the stator to establish a rotating magnetic field, the ratio of reactance to resistance of one winding is made different from the other. By doing this, both windings carry currents which are out-of-phase with each other (this is the meaning of split-phase). The magnetic fields produced by these currents create the necessary torque on the rotor to make the motor self-starting when connected to the single-phase supply. Two motors operating on this principle are called: (i) *resistance split-phase motor*, and (ii) *capacitor split-phase motor*. These will now be discussed.

Resistance split-phase motor In this motor the starting winding is designed to have more resistance in it than the running winding – this can often be seen by looking at the windings inside the stator slots of a typical motor and noting their thickness, the starting winding is much thinner. The resistance of the running winding is, however, low in comparison with its inductance. When it is connected to the supply, current through it will lag behind the current flowing through the starting winding and a rotating field will be established. In order to reduce I^2R losses, the starting winding is arranged to be disconnected when the rotor has reached between 75% and 80% of its full load speed. The method commonly used to achieve disconnection is a centrifugal switch mechanism attached to the rotor shaft. Figure 4.7 is a typical wiring circuit diagram and Figure 4.8(b) shows a rotor fitted with a centrifugal switch. The starting torque of this motor is approximately twice its full load torque.

Capacitor split-phase motor The starting winding of this motor incorporates a capacitor instead of additional resistance. When the motor is first switched on, current in the starting winding will lead the current in the running winding by approximately 90° (elect). This gives the motor a better starting performance than the previous motor. A typical motor is shown in Figure 4.8(a) with its capacitor mounted on the casing. A wiring diagram is also shown in Figure 4.7(b). It will be seen that a

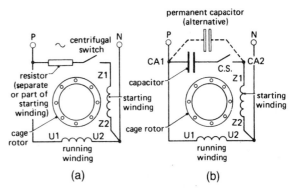

Fig. 4.7 Types of split-phase induction motor (a) resistance start (b) capacitor start.

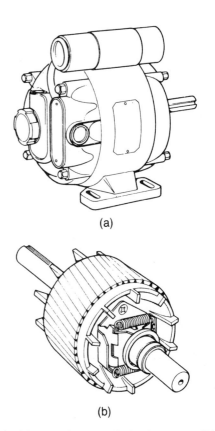

Fig. 4.8 (a) Capacitor start induction motor (b) cage rotor showing centrifugal mechanism.

permanent capacitor may alternatively be wired in series with the starting winding. This is done to improve the power factor of the circuit caused by the lagging current created by the stator windings – often a problem with all a.c. motors. The application of these motors is generally limited to small output ratings below 750 W.

(b) Shaded-pole motors

There are two types of shaded-pole motor, both of which are forms of induction motor that operate on a phase shift principle. The one shown in Figure 4.9 is described as a 'unicoil' type, having a wound coil or bobbin for its main stator winding. On its laminated stator will be seen two sets of copper shading rings which are magnetically displaced to create an artificial phase shift. The other type of motor is called a 'salient-pole' motor, having prominent stator poles similar to those of a d.c. machine. Each of its poles is fitted with a copper shading ring as previously mentioned. These rings are really short-circuit coils which serve the purpose of delaying the magnetic flux as it passes around the stator. Figure 4.10 illustrates this.

When the supply is switched on, eddy currents are induced in the shading ring which then sets up its own magnetic field. This flux opposes the main flux and causes it to concentrate on the opposite side of the pole piece. This action causes a slight delay before the flux in the shaded area reaches a maximum value. When the main flux starts to decrease, the field inside the shading ring increases, giving the effect of sweeping from one side of the pole to the other. This shift is sufficient to cause the rotor to move, making it self-starting. Unicoil motors are smaller in output rating than the salient-pole types, being between 2 W and 21 W. They are used mainly for oven fans and timers. Salient-pole motors, particularly the four-pole types, have output ratings of the order of 125 W; their higher efficiencies make them ideally suited for washing machines.

Commutator motors

(a) Universal motor

This motor resembles the d.c. series motor in its construction and operation. Briefly, it consists of a

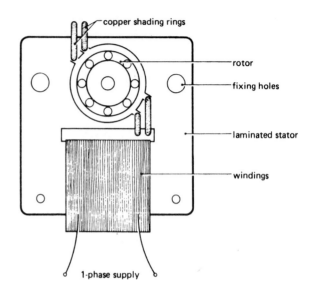

Fig. 4.9   Shaded-pole (unicoil) motor.

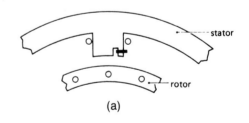

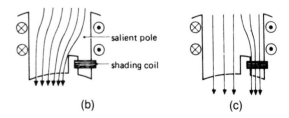

Fig. 4.10   (a) Salient-pole stator (b) Effects of shading coil opposing main field (c) Shading coil aiding main field.

fixed series field winding on a laminated yoke, often distributed around in a similar manner to the motor described earlier. The motor has an armature and commutator which is connected in series with the field coils.

Its operation principle is based upon the interaction of two fields produced by the same current. The main series field requires little explanation, it produces an alternating flux as before, but the second magnetic field is produced by passing current into armature conductors via a *commutator*. The interaction of these two fields is shown in Figure 4.11. The diagram is simplified, showing how force is exerted in opposite directions when current passes around a single armature conductor. Because the supply is alternating, both field polarity and armature conductor polarity will change at the same time instants and the motor will continue to run in the same direction. The a.c. supply will however set up eddy currents, causing I^2R losses and overheating. To reduce this effect the whole magnetic circuit must be laminated.

Another problem with this motor when it is used on a.c. is that its armature coils are subject to the effects of inductance which causes sparking at the brushes. This is overcome by arranging a neutralizing or compensating coil at right-angles to the field winding and connecting it in series with the armature. Figure 4.12 is a typical circuit for this motor; also shown is the motor speed–torque characteristic and power factor–torque characteristic.

The majority of these motors are used in vacuum cleaners, hand mixers, hair dryers, etc., and they are capable of running at very high speeds compared with induction motors (a cylinder vacuum cleaner may run at 300 rev/s). One of their disadvantages is that the brushes on the commutator are prone to wear – especially when running at high speeds for long periods.

(b) Repulsion motor

This type of motor is an induction form of series motor. It has a commutator type of armature similar to the universal motor, but its brushes are short-circuited by a heavy conductor of low resistance. A circuit diagram of this motor is shown in Figure 4.13.

In terms of its operation, the stator winding

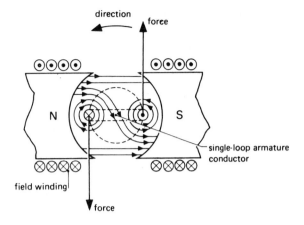

Fig. 4.11 Field interaction of universal series motor showing armature direction.

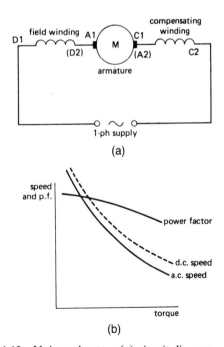

Fig. 4.12 Universal motor (a) circuit diagram (b) Speed-torque and power factor-torque characteristics.

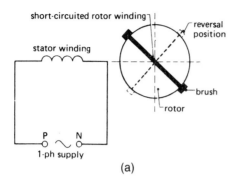

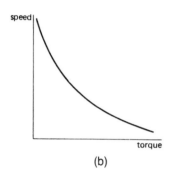

Fig. 4.13 Repulsion motor (a) circuit diagram (b) Speed-torque characteristics.

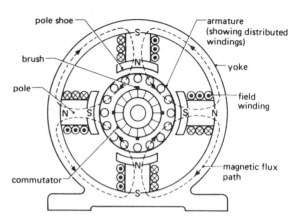

Fig. 4.14 Cross section of a four-pole d.c. machine showing basic construction and magentic flux paths.

produces a current in the rotor winding by induction – in much the same way as transformer action. Both stator and rotor magnetic fluxes are designed to interact to produce a torque necessary for running. However, before this can happen, the brushes must be set at some position other than in line with the stator field or at 90° to it. The preferred position is somewhere between these two positions (in practice about 20°). If the brushes are set in line with the stator, both stator field and rotor field act in opposition with each other and no repulsion will take place. In the other extreme position (i.e. when the brushes are at right-angles to the stator field) then the e.m.f.s induced in each half of the rotor winding cancel out, again giving rise to no movement on the rotor.

The motor has a series motor speed-torque characteristic and it may be provided with a compensating winding to improve its communication – i.e. to stop unnecessary sparking. Its starting torque

is good and its speed can be controlled by altering the brush position.

Two other types of repulsion motor are: (i) the *repulsion–induction motor* and (ii) *the repulsion-start induction motor*. Briefly, the first type has a cage winding embedded under its armature winding which gives the motor a constant speed up to full load. The second type is fitted with a centrifugal switch which operates at about 75% of the normal full-load speed. The switch short-circuits the commutator segments and lifts the brushes; it then operates as a normal induction motor.

Direct-current motors

Basic construction

These motors, as mentioned earlier, are named according to their field system arrangement. They are called the *series motor, shunt motor* and *compound motor* – the latter having both series and shunt fields. Their basic constructural details are the same and a typical machine is illustrated in Figure 4.14. Like all other motors, the two essential parts are the *yoke* (or frame) and rotor (called the armature). The yoke, the stationary part, is constructed of cast steel or fabricated mild steel having good magnetic properties. Attached to the yoke are *pole cores* which are usually made from mild steel plates riveted together.

Some machines may have their poles solidly cast with the yoke. Fitted to these are laminated *pole shoes* designed to support the *field coils* and also increase the cross-sectional area of the poles – this is to reduce the reluctance of the air gap between poles and armature. The field coils are concentrated windings wound directly on the poles. They may be either circular or rectangular in section and insulated. Large series field coils may be formed from copper strip, whereas shunt field coils are noticeably smaller in cross-sectional area and generally contain more turns.

The *armature* consists of iron laminations which are insulated from each other and assembled on a mild steel shaft. The laminations are slotted to accommodate the armature winding (sometimes called commutator winding). The armature winding is distributed around the armature making regular connections on to a *commutator* (see Figure 4.15). Armature windings are usually divided into two distinct types, namely *lap winding* and *wave winding*. The former type has its connections taken to adjacent commutator segments, whereas in the latter case connections are taken to segments some distance apart. It will be seen that the commutator is made up of copper segments insulated from each other by mica. Connections to the commutator are made by fixed carbon or graphite brushes in associated brushholders. Finally, the armature is mounted on bearings which are themselves supported by end plates.

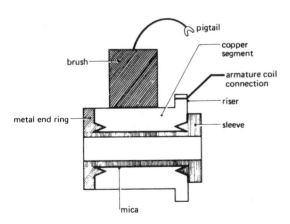

Fig. 4.15 Cross section of a commutator showing copper segments insulated by mica.

Principles of operation

Direct current motor operation is based again on the interaction between two magnetic fields, in this case established by the main field windings and armature windings. A *torque* results, causing movement of the armature (Figure 4.11). It is seen that the magnetic field passing across the armature is distorted. Each conductor on the armature will experience a force in this manner. Torque (T) therefore, can be regarded as the *force* resulting in rotary movement multiplied by the *radius* at which it acts. It is expressed in newton-metres (Nm). In order to understand its relationship with magnetic flux and armature current, one needs to consider the force acting on a current-carrying conductor situated in a magnetic field. This is given by the expression:

$$F = BIl \text{ newtons} \tag{2.16}$$

where B is the magnetic flux density in teslas
I is the current in amperes
l is the conductor length in metres at right-angles to the magnetic field

In the electric motor, the magnetic flux is produced by the main poles, i.e. the flux per pole (Φ); the current carried by the conductor is the armature current (I_a), and the conductor length is fixed for a given armature. Expression (2.16) can now be arranged to read:

$$F \propto \Phi I_a \tag{4.3}$$

but since torque is proportional to force ($T \propto F$), then

$$T \propto \Phi I_a \text{ newton-metres} \tag{4.4}$$

This is a very important expression since it enables one to find out how performance can be changed and what variables need to be changed in order to achieve it. Once the torque is found, the mechanical output power can be found. This is expressed as $P = 2\pi nT$ **watts** (where n is the speed in revolutions per second).

When the armature rotates, the conductors on it have e.m.f.s induced in them. These induced e.m.f.s act in opposition to the current taken by the motor and also to the applied voltage (V), obeying Lenz's law. The name given to this opposing voltage is called *back e.m.f.* (E). It will always be less than the applied voltage, but it is important because it controls the amount of current through the armature when the motor is running.

When a motor is first switched on, it requires a starting resistance to limit the inrush of current to its armature. As the motor picks up speed, the back e.m.f. rises and automatically limits the current. The starting resistance can then be cut out of circuit. The relationship between the armature current, back e.m.f. and armature resistance may be expressed as:

$$V = E + I_a R_a \text{ volts} \qquad (4.5)$$

Since the back e.m.f. is a generated voltage (induced), it is proportional to the speed of the armature and also to the magnetic flux per pole.

Thus $\quad E \propto \Phi n \qquad\qquad\qquad (4.6)$

Series motor

This motor has its field winding connected in series with the armature. This means that both field and armature have the load current flowing through them. From expression (4.4), the torque, in this case, is proportional to the square of the current ($T \propto I^2$) – i.e. until magnetic saturation is reached. If the motor is started under heavy load it will demand a large current from the supply and a large torque will be created – for a doubling of I_a the torque will increase four times. If, however, the load is reduced, the magnetic field becomes weak and the speed increases. If the load is completely thrown off, its speed is likely to become dangerously high, it should therefore always be connected to some form of load.

The speed may be varied by using a diverter resistance connected in parallel with the field winding. A diagram of this arrangement is shown in Figure 4.16 together with typical speed–torque characteristics. It should be pointed out that it is possible to vary the speed using a diverter across the armature.

D.C. series motors are used where high starting torques are required, e.g. hosts, fans and traction work.

Shunt motor

This motor has its field winding connected in parallel with the armature. This means that the field winding receives full supply voltage and so the strength of the magnetic field is approximately constant at all loads. The torque of this motor is proportional to the armature current and its speed is fairly constant – falling slightly at high loads because of the effects of armature reaction.

Generally speaking, the speed of the motor can be varied by controlling either its field current or its armature voltage. By controlling the armature voltage with a fixed field excitation, the motor is capable of developing full load torque over its entire speed range. The output power will then increase with speed. Control of the motor field, which usually occurs when the armature voltage is at the maximum rated value, will give a constant output power characteristic. As the field current is reduced, speed will increase and its torque will be reduced. One method of controlling the field current is to place a variable resistor (field regulator) in series with the

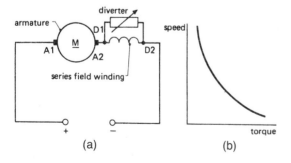

Fig. 4.16 (a) Circuit diagram of d.c. series motor (b) Speed-torque characteristics.

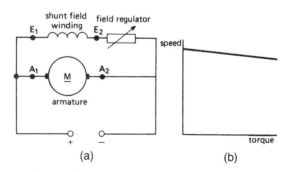

Fig. 4.17 (a) Circuit diagram of d.c. shunt motor (b) Speed-torque characteristics.

shunt winding. Figure 4.17 is a typical shunt motor circuit; also illustrated are the speed and torque characteristics.

Because of the motor's fairly constant speed, it is used for a number of industrial drives, such as machine tool drives, compressors and pumps. Applications requiring a motor with a constant torque are conveyor lines, crane hoists and milling machines.

Compound motor

This motor has series and shunt field windings. In practice, one winding provides the main field and the other an auxiliary field. When they are connected so that their magnetic fields *assist* each other, the motor is called a *cumulative compound motor*. When the magnetic fields are made to *oppose* each other, the motor is called a *differential compound motor*. The motor's connections and characteristics are shown in Figure 4.18.

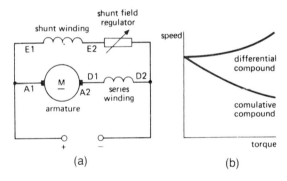

Fig. 4.18 (a) Circuit diagram of d.c. compound motor (b) Speed-torque characteristics.

Generally speaking, the differential compound ,motor is hardly used, its speed at low torque is limited by the shunt winding and on heavy loads it will tend to race away. This restricts its application for loads requiring constant speed.

In the cumulative compound motor, the magnetic field on no-load is due to the shunt winding (being the chief winding) and the motor behaves like a shunt motor having a constant speed. As load is placed on the motor, the series field winding increases in strength giving a series-motor characteristic, i.e.

speed falling as torque increases. It is used where large starting torques are required or where the supply is subject to fluctuations, e.g. traction systems, haulage gear, rolling mills, trolley buses.

Motor reversal

Three-phase induction motor: change any two supply leads.

Single-phase, split-phase induction motor: change leads of any one winding.

Single-phase, shaded-pole motor: difficult, shading rings need changing to other side of salient pole.

Single-phase, universal motor: change either stator or armature connections.

Single-phase, repulsion motor: movement of brushes around commutator or by fitting motor with two windings.

Direct-current motors: change either field windings or armature connections, not both.

Motor calculations

Example 1

Determine the synchronous speed (n_s) and rotor speed (n_r) from the following induction motor data:

a) $f = 50$, $p = 2$ and $s = 0.03$
b) $f = 50$, $p = 4$ and $s = 0.03$
c) $f = 50$, $p = 6$ and $s = 0.03$
d) $f = 50$, $p = 6$ and $s = 0.05$
e) $f = 60$, $p = 6$ and $s = 0.05$

Note: f is the frequency of supply in hertz
       p is the number of stator poles
       s is the slip in per unit values

Solution

a) $n_s = \dfrac{f}{p}$ $n_r = n_s(1 - s)$

    $= \dfrac{50}{1}$ $= 50(1 - 0.03)$

    $= \underline{50 \text{ rev/s}}$ $= \underline{48.5 \text{ rev/s}}$

b) $n_s = \dfrac{50}{2}$ $n_r = 25(1 - 0.03)$

    $= \underline{25 \text{ rev/s}}$ $= \underline{24.25 \text{ rev/s}}$

c) $n_s = \dfrac{50}{3}$

$= 16.66$ rev/s

d) $n_s = \dfrac{50}{3}$

$= 16.66$ rev/s

e) $n_s = \dfrac{60}{3}$

$= 20$ rev/s

$n_r = 16.66(1 - 0.03)$

$= 16.17$ rev/s

$n_r = 16.66(1 - 0.05)$

$= 15.83$ rev/s

$n_r = 20(1 - 0.05)$

$= 19$ rev/s

Example 2

Determine the power output, power factor and efficiency of a three-phase motor having the following test data:

1 speed 23.75 rev/s
2 input wattmeter reading 16 920 W
3 voltmeter reading 400 V
4 ammeter reading 39.5 A
5 brake pulley diameter 0.33 m
6 effective pull at circumference of pulley 564.44 N

Solution

Power output (mechanical) is given by the expression:

$P_o = 2\pi \, nT$ watts
$= 2 \times 3.142 \times 23.75 \times (564.44 \times 0.165)$
$= 13\ 899$ W or 13.9 kW

Note: $T = F \times$ radius

Efficiency $= \dfrac{\text{output}}{\text{input}}$

$= \dfrac{13\ 899}{16\ 920}$

$= 0.82$ p.u. (82%)

Power input (electrical) is given by the expression:

$P_i = \sqrt{3}V_L I_L \cos \phi$ watts

where $\cos \phi$ the power factor
By transposition:

$\cos \phi = \dfrac{P_i}{\sqrt{3}V_L I_L}$

$= \dfrac{16\ 920}{1.732 \times 400 \times 39.5}$

$= 0.62$ lagging

Example 3

a) What is an interpole? Explain where it is connected in a machine.
b) Draw a diagram of a d.c. machine's main salient pole showing both series and shunt field connections to give cumulative compound characteristics. Show current directions in the field windings and pole polarity.

Solution

a) An interpole is a small pole fitted between the main poles of a d.c. machine in order to neutralize the effects of armature reaction which causes sparking at the brushes. Interpoles are connected in series with the armature connections.
b) See Figure 4.19, which is a main pole, not an interpole.

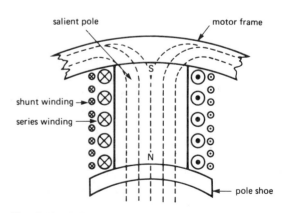

Fig. 4.19 Salient electromagnetic pole.

Example 4

The speed of a 220 V d.c. motor with an armature current of 10 A and armature resistance 0.5 Ω is 12.5 rev/s. What would be its speed if the armature current were increased to 30 A? Assume the armature voltage (applied voltage) and field current remain unchanged.

Solution

Since $E = V - I_aR_a$

where E is the back e.m.f.
      V is the applied volts
      I_a is the armature current
      R_a is the armature resistance

First condition:

$$E_1 = 220 - (10 \times 0.5) = \underline{215\ V}$$

Second condition:

$$E_2 = 220 - (30 \times 0.5) = \underline{205\ V}$$

The back e.m.f. is proportional to the speed of the armature and magnetic flux per pole. If the field current remains unchanged, the conditions can be represented by the expression:

$$E \propto \phi n$$

Thus $E_1 \propto n_1$
and $E_2 \propto n_2$

By proportion $n_2 = \dfrac{n_1 \times E_2}{E_1} = \dfrac{12.5 \times 205}{215}$
$$= \underline{11.9\ rev/s}$$

Example 5

The current taken by a 240 V, 50 Hz single-phase induction motor is 39 A at a power factor lagging of 0.75. Determine:
a)   the input power in kilowatts
b)   the kilovoltamperes

Solution

a $P = VI\cos \phi$
    $= 240 \times 39 \times 0.75$
    $= 7020\ W$
    $= \underline{7.02\ kW}$

b) $VI = 240 \times 39$
    $= 9360\ VA$
    $= \underline{9.36\ kVA}$

See Figure 4.20.

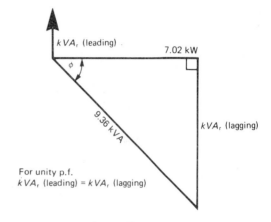

Fig. 4.20 Power factor diagram.

Example 6

a)   Explain with the aid of diagrams the method of reversing the direction of rotation of *each* of the following types of motor:
   (i)   three-phase induction
   (ii) single-phase capacitor start
   (iii) d.c. shunt
   (iv) series

b)   Calculate the full-load current of a 48 kW 3-phase 415 V motor, given that the efficiency and power factor at full load are 85% and 0.9 respectively.

Solution

a)   Methods of reversing the direction of rotation of motors are given on page 74 of *Electrical*

Installation Technology 2.

    (i)   Change any two supply leads.

    (ii)   Change connections of running winding *or* starting winding.

    (iii) Change connections of shunt field winding *or* armature winding.

    (iv) Change connections of series field winding *or* armature winding.

    See Figure 4.21.

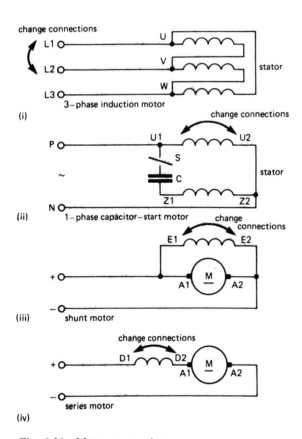

Fig. 4.21 Motor connections.

Example 7

A 1 kW, 240 V, 50 Hz, 2-pole single-phase, induction motor operates with 5% slip, 75% efficiency and 0.7 power factor on full load.

a)   Draw a labelled circuit diagram of a push button starter with undervoltage and overcurrent protection for the above motor.

b)   For full-load conditions calculate the:

    (i)   input power and current

    (ii) motor speed   CGLI/II/83

Solution

a)   See Figure 4.22

b)   (i) Since efficiency $= \dfrac{\text{output}}{\text{input}}$

$$\text{then input} = \frac{\text{output}}{\text{efficiency (p.u.)}}$$

$$= \frac{1\,000}{0.75}$$

$$= 1.33 \text{ kW}$$

because input $= V_P I_P \cos \phi$

By transposition:

$$I = \frac{P}{V_P \cos \phi}$$

$$= \frac{1333.3}{240 \times 0.7}$$

$$= \underline{7.94 \text{ A}}$$

(ii) Since $n_s = \dfrac{f}{p}$

$$= \frac{50}{1}$$

$$= \underline{50 \text{ rev/s}}$$

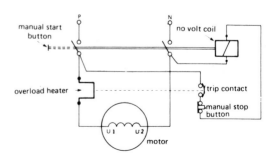

Fig. 4.22 Single-phase a.c. motor.

where n_s is the synchronous speed
f is the supply frequency
p is the number of pairs of poles

since $\quad\quad n_r = n_s(1 - s)$
where n_r is the rotor speed

Then: $\quad\quad n_r = 50(1 - 0.05)$
$\quad\quad\quad\quad\quad = \underline{47.5 \text{ rev/s}}$

Example 8

a) Explain briefly how the back e.m.f. and the current change during the starting of a d.c. motor.

b) A 200 V shunt-wound motor has an armature resistance of 0.25 Ω and a field resistance of 200 Ω. The motor gives an output of 4 kW at an efficiency of 80%. For this load calculate:
   (i) the motor power input in kW
   (ii) the load current
   (iii) the motor field current
   (iv) the armature current
   (v) the back e.m.f.

c) What are the IEE requirements for the rating of fuses protecting a circuit feeding a motor?

$\quad\quad\quad\quad\quad\quad\quad\quad\quad\quad$ CGLI/II/81

Solution

a) The back e.m.f. of a d.c. motor is a generated e.m.f. created by the armature conductors cutting the main field as the armature revolves. This induced e.m.f. acts in opposition to the supply voltage and is always less than the supply voltage – the difference being the armature voltage drop.

As the armature accelerates, its back e.m.f. increases and the armature current decreases. Also, as the load current increases, the back e.m.f. decreases. If the armature stops revolving there will be no back e.m.f. and with the supply switched on, excessive current will be drawn by the motor.

b) $\quad$ (i) Since $\quad$ efficiency $= \dfrac{\text{output}}{\text{input}}$

$$\text{input} = \dfrac{\text{output}}{\text{efficiency (p.u.)}}$$

$$= \dfrac{4000}{0.8}$$

$$= 5000 \text{ W}$$

$$= \underline{5 \text{ kW}}$$

(ii) $\quad\quad I_L = \dfrac{\text{power input}}{\text{supply voltage}}$

$$= \dfrac{5000}{200}$$

$$= \underline{25 \text{ A}}$$

(iii) $\quad\quad I_f = \dfrac{\text{supply voltage}}{\text{field resistance}}$

$$= \dfrac{200}{200}$$

$$= \underline{1 \text{ A}}$$

(iv) Since $\quad\quad I_L = I_a + I_f$

then $\quad\quad I_a = I_L - I_f$

$$= 25 - 1$$

$$= \underline{24 \text{ A}}$$

(v) $\quad\quad\quad\quad E_b = V - I_a R_a$

$$= 200 - (24 \times 0.25)$$

$$= \underline{194 \text{ V}}$$

c) Reference should be made to the IEE Wiring Regulations (15th Edition), Section 434, Reg. 434–5. In the 14th Edition, Reg. A–68 allowed fuses to be rated up to twice that of the cables between the fuse and the starter, provided the starter afforded overload protection. However, the new requirements allow the use of an overload device complying with Section 433 to protect the conductors on the load side of the device, provided that it has a rated breaking capacity not less than the prospective short-circuit current at the point of installation.

Note: While it is not required in the question, Figure 4.23 is a diagram of a shunt motor showing some of the terms mentioned in the solution above.

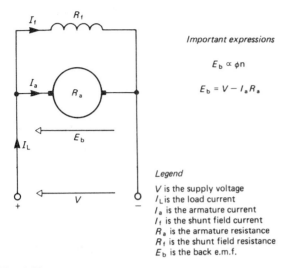

Important expressions

$$E_b \propto \phi n$$

$$E_b = V - I_a R_a$$

Legend

V is the supply voltage
I_L is the load current
I_a is the armature current
I_f is the shunt field current
R_a is the armature resistance
R_f is the shunt field resistance
E_b is the back e.m.f.

Fig. 4.23

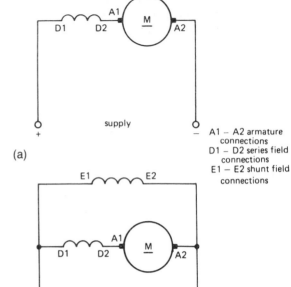

(a)

(b)

Fig. 4.24

A1 – A2 armature
connections
D1 – D2 series field
connections
E1 – E2 shunt field
connections

Example 9

a) Describe briefly the following parts of a d.c. motor:
   (i) armature
   (ii) commutator
   (iii) field system
b) Draw a circuit diagram for:
   (i) a d.c. series motor
   (ii) a d.c. compound motor

CGLI/II/82

Solution

a) (i) The armature is the name for the rotating part of the machine and is made up of many laminations of soft-magnetic-alloy material into which armature coils are assembled.

(ii) The commutator is part of the armature, serving the purpose of transferring an external current to the armature conductors via brush-gear.

(iii) The field system is that part of the d.c. motor which produces the excitation flux. It comprises the main poles and field windings which identify the machines as a *series, shunt* or *compound motor*.

b) (i) See Figure 4.24(a)
   (ii) See Figure 4.24(b)

Example 10

A 6-pole, 415 V, 50 Hz, three-phase induction motor operates at 0.7 power factor lagging and drives and elevator lifting 100 kg at a rate of 3.6 m/sec. If the elevator has an efficiency of 75% and the motor an efficiency of 85%, determine for full load conditions:
a) the motor's output and input
b) the motor's line current and phase current assuming the windings are delta connected
c) the motor's synchronous speed and rotor speed assuming a 0.05 p.u. slip

Solution

a) Work done/second = force × distance
                     = 100 × 3.6
                     = 360 kgf

If 1 kgf = 9.81 N, then power required by elevator is:

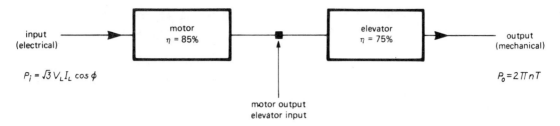

Fig. 4.25 Efficiency of machines.

$9.81 \times 360 = 3531.6$ W

Motor's output = elevator's input (see Figure 4.25)

$$= \frac{3531.6 \times 100}{75}$$

$$= \underline{4708.8 \text{ W}}$$

$$\text{Motor's input} = \frac{4708.8 \times 100}{85}$$

$$= \underline{5539.76 \text{ W}}$$

b) Since $\qquad P = \sqrt{3}V_L I_L \cos \phi$

then $\qquad I_L = \dfrac{P}{\sqrt{3}V_L \cos \phi}$

$$= \frac{5539.76}{\sqrt{3} \times 415 \times 0.7}$$

$$= \underline{10.13 \text{ A}}$$

$$I_P = \frac{I_L}{\sqrt{3}}$$

$$= \frac{10.13}{\sqrt{3}}$$

$$= \underline{5.85 \text{ A}}$$

c) $\qquad n_s = \dfrac{f}{p}$

$$= \frac{50}{3}$$

$$= \underline{16.66 \text{ rev/s}}$$

$$n_r = n_s(1 - s)$$

$$= 16.66(1 - 0.05)$$

$$= \underline{15.83 \text{ rev/s}}$$

Exercise 4

1 From your knowledge of sinewaves, construct phasor diagrams for the six positions shown in Figure 4.26. This figure is typical of the m.m.f.s necessary to produce rotation in a capacitor-start, single-phase, split-phase induction motor.

2 *a)* A 240 V single-phase motor takes a current of 7 A and has a power input of 1.1 kW. Calculate the
  (i) input kVA
  (ii) power factor
  (iii) power output if the efficiency is 72%

  b) State the requirements of the *IEE Regulations* with respect to the minimum size of cable supplying a motor.

  c) State *two* typical applications where the requirement of Regulation 552–4
  (i) does apply
  (ii) does not apply

CGLI/II/1987

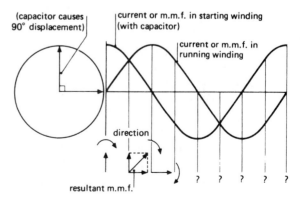

Fig. 4.26 Rotating field of split-phase capacitor induction motor.

3 Explain the meaning of the following terms associated with motors:
   a) air-gap
   b) armature
   c) commutator
   d) back e.m.f.
   e) synchronous speed

4 a) Draw a circuit diagram of a single-phase, split-phase, cage induction motor with a capacitor connected to improve the power factor of the circuit.
   b) A 250 V, 4 kW, single-phase motor has an efficiency of 80% and a power factor of 0.8 lagging when operating at full load. Calculate
      a) the input power
      b) the input current
      c) The kVAr rating of a capacitor required to improve the power factor to unity.
                                    CGLI/II/1986

5 Draw a labelled circuit diagram of a 415 V, three-phase motor controlled by a direct-on-line, pushbutton-operated, contactor starter. Include in the circuit all the equipment and devices necessary to comply with the following requirements:
   a) means to prevent automatic restarting after a stoppage
   b) isolation of the circuit
   c) switching for mechanical maintenance of the motor
   d) emergency switching including a remote position
   e) overload and short circuit protection for motor and cables.
                                    CGLI/II/1986

6 a) Describe with the aid of sketches the operation of the following types of overload device:
      (i)   oil-dashpot
      (ii) bimetallic
   b) A pump and a single-phase 240 V, 50 Hz motor have a combined efficiency of 76% with a motor power factor of 0.85 lagging. If the output power of the pump is 4 kW, calculate the motor
      (i)   input power
      (ii) line current.
                                    CGLI//II/1984

7 a) Explain why it is necessary to have a resistance starter for the safe starting of a d.c. motor.
   b) Draw a circuit diagram of the internal connections of d.c. shunt motor face-plate starter and explain the operation of the no-volt release and overload release.

8 Explain with the aid of a diagram *two* methods of controlling the speed of a d.c. shunt motor. Show the motor's speed–torque characteristics.

9 A 240 V d.c. shunt motor has an armature resistance of 0.2 Ω and a field winding resistance of 120 Ω. If the current at no-load is 5 A and the speed is 30 rev/s, determine the motor's speed at a full-load current of 40 A.

10 a) In your own words, try to explain the operation of a three-phase cage induction motor.
   b) Why is it not possible for the rotor to travel at the same speed as the rotating synchronous speed?

CHAPTER FIVE

Transformers and rectifying devices

After reading this chapter you will be able to:

1 Describe the operation of single-phase transformers.

2 Draw diagrams of core-type and shell-type double-wound transformers and explain why they are laminated.

3 Perform calculations associated with transformation ratios and efficiency.

4 State the losses occurring in transformers and know why they need cooling when on load.

5 Describe the basic theory associated with semiconductor rectifiers such as the diode and thyristor.

6 State the difference between forward and reverse bias of junction diodes.

7 Draw circuit diagrams of half-wave and full-wave rectification and sketch their output signals.

8 Draw the graph of a thyristor's output current after its gate has been triggered.

Transformers

A transformer is an a.c. electrical apparatus which takes in power at one voltage and delivers it at another voltage. It operates on the principle of electromagnetic induction which was discussed in Chapter 2.

If a coil is connected to a d.c. source of supply, the current passing through it would produce a magnetic field with a definite north and south pole at either end. If the coil were connected to an a.c. supply, the changing current would produce a changing magnetic field and this would have a changing north and south pole. By its own *self-inductance* a voltage would be created. *Autotransformers, variacs* and *ring-type transformers* all have one winding and it is a question of where the supply is fed into this winding or what tappings are made on it which dictates the magnitude of voltage that is required on the output side. The winding section of the transformer connected to the a.c. supply is called the *primary* and the section connected to output load is called the *secondary* (see Figure 5.1).

If a second coil were placed next to the a.c.-connected coil as described above, the first coil would be called the primary winding and the second coil called the secondary winding. Despite there being no

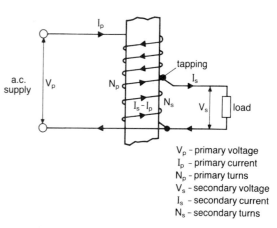

V_p – primary voltage
I_p – primary current
N_p – primary turns
V_s – secondary voltage
I_s – secondary current
N_s – secondary turns

Fig. 5.1 Current flow in windings of an auto transformer.

electrical connection between the coils, flux linkages from the first coil would cut through the secondary winding and it will become induced, creating a voltage by *mutual inductance*. The closer the coils are the greater will be the induced e.m.f. and this voltage can be made much greater by using metal between the two coils to concentrate the magnetic flux linkages.

Transformers which are designed with two or three windings are called *double-wound transformers* and they use laminated steel cores to support their windings. The reason for using laminated cores and not solid metal cores is because of *induced eddy currents*. These are circulating currents set up in the metal which cause unnecessary heating and power losses. Figure 5.2 illustrates the path they take. If the two cores had the same resistance (say 5 Ω) and the circulating current was say 5 A, the power loss in diagram (a) would be $P = I^2R = 25 \times 5 = 125$ W. In the case of diagram (b), the laminations are lightly insulated from each other and the circulating current in each is only 1 A. The loss in each lamination is $P = I^2R = 1$ W and for the whole laminated core it is 5 W. This should help to explain why it is necessary to keep all a.c. magnetic systems highly laminated.

Figure 5.3 shows a typical diagram of a double-wound transformer and it will be seen that the supply primary winding will circulate an alternating magnetic flux in the core. This flux (assuming no leakage) will link with the secondary winding and an e.m.f. will be induced in it. The e.m.f. induced will depend on the number of turns on the secondary winding as well as the rate of change of the magnetic flux but since both windings are linked by the same flux, their induced

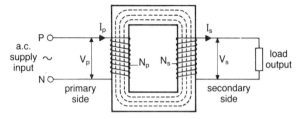

Fig. 5.3 Double-wound transformers.

e.m.f.s are proportional to the number of turns in each coil. In theory, the no-load current of a transformer is very small, creating negligible voltage drop and, as a result of this, the induced e.m.f. in both windings is almost equal to the terminal voltages. These winding turns (N) and terminal voltages (V) can be expressed as a ratio:

$$\frac{V_p}{V_s} = \frac{N_p}{N_s} \tag{5.1}$$

where V_p is the primary voltage
V_s is the secondary voltage
N_p is the primary winding turns
N_s is the secondary winding turns

The expression (5.1) means that the voltage ratio is the same as turns ratio.

Generally speaking, because a transformer has no moving parts it is a highly efficient piece of static plant and therefore its efficiency is around 98%. The chief losses in a transformer are:

1 iron losses – due to eddy currents and hysteresis
2 copper losses – I^2R losses in the windings

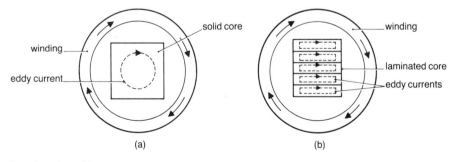

Fig. 5.2 Paths taken by eddy currents.

This implies that its output power (P_o) is almost equal to its input power (P_i). This can be written as:

$$V_p\, I_p \cos \phi \simeq V_s\, I_s \cos \phi$$

and since the power factors are almost equal then

$$V_p\, I_p = V_s\, I_s$$

Expressing (5.2) as a ratio:

$$V_p/V_s = I_s/I_p \tag{5.2}$$

Thus the current ratio is inversely proportional to the turns ratio. If (5.1) and (5.2) are combined, then the transformation ratio is:

$$\frac{V_p}{V_s} = \frac{N_p}{N_p} = \frac{I_s}{I_p}$$

One of the difficult things to understand is what happens when the transformer is connected to a load, when its secondary winding carries current. This secondary current creates another magnetic flux in the core which tries to cancel out the flux produced by the current in the primary winding. To restore the magnetic flux in the core the primary winding has to draw a current from the supply so that the e.m.f.s of both windings remain unchanged. This leaves the transformer with a *constant flux* at all loads and it also means that the transformer's *iron losses* will be *constant* at all loads.

From a constructional point of view, Figure 5.4 shows the normal methods of designing double-wound transformers; (a) is a core type having two limbs and (b) is a shell type having three limbs with the middle limb being twice the size of the outer limbs. The windings are either of the concentric type or sandwich type as shown in Figure 5.14(c). Attention should be paid to the low-voltage and high-voltage windings, since it is practice to place the low-voltage winding next to the core.

A three-phase transformer delta-star connected transformer is shown in Figure 5.5. It is a 11 kV step-down transformer used for supplying 415 V/240 V to a three-phase, 4-wire distribution system. Note the lettering of its primary and secondary windings.

Supply transformers are normally classified according to their grouping, i.e. the phase relationship at their terminals. A star–star or delta–delta transformer will have no phase displacement

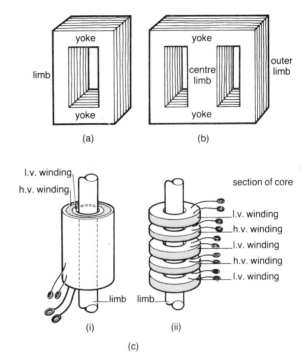

Fig. 5.4 Transformer cores.

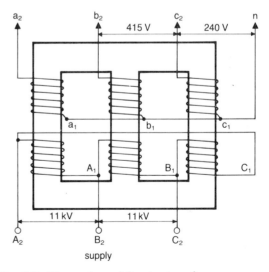

Fig. 5.5 Three-phase delta-star transformer.

and is called a Group Y-y-0 or D-d-0. The delta–star transformer has a phase shift between its high- and low-voltage windings of $\pm30°$ and it is called a Group

D-y-1. There are several other groupings but these will not be discussed. These groupings are of real importance, particularly when transformers are to be connected in parallel to share a load.

Mention has already been made of transformer losses and these are evident when on load, where the power losses are converted into heat. Since the transformer is static plant, natural air cooling may not be sufficient. Figure 5.6 shows one method of overcoming this problem using an oil-filled transformer. Its construction makes use of cooling tubes to allow it to convert heated oil away from the main core. There are several other methods of cooling such air-blast and water cooling but these will not be discussed.

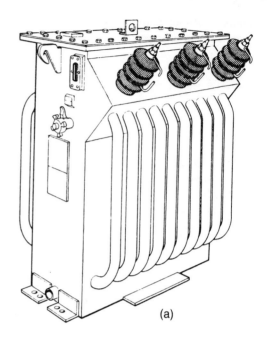

(a)

Transformer calculations

1   A double-wound, single-phase step-down transformer having a transformation ratio of 20:1, has a primary voltage of 6600 V and feeds a load of 13.2 kVA. Assuming the losses to be negligible, determine the secondary voltage and secondary and primary currents.

Solution

From expression (5.3),

$$\frac{V_p}{V_s} = 20 \quad \frac{I_s}{I_p} = 20 \quad \text{and} \quad \frac{N_p}{N_s} = 20$$

Therefore

$$V_s = \frac{V_p}{20} = \frac{6600}{20}$$

$$= \underline{330\ V}$$

Since

$$P_{\text{input}} = P_{\text{output}}$$
$$VI_{\text{input}} = VI_{\text{output}}$$

$$6600 \times I_p = 13\ 200\ \text{VA}$$

Therefore

$$I_p = \frac{13\ 200}{6600}$$

$$= \underline{2\ A}$$

And

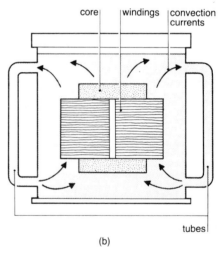

Fig. 5.6   Three-phase oil immersed transformer.

$$I_s = \frac{13\ 200}{330} \quad \text{i.e.} \quad I_s = \frac{VI_{\text{output}}}{V_{\text{secondary}}}$$

$$= \underline{40\ A}$$

Alternatively,

$$I_s = I_p \times \text{ratio}$$
$$= 2 \times 20$$
$$= \underline{40\ A}$$

If we were told that the transformer had a volt per turn of 0.1, then the number of turns on the primary would be:

$$N_p = \frac{\text{primary volts}}{\text{volts per turn}} = \frac{6600}{0.1}$$

$$= 66\ 000 \text{ turns}$$

Similarly

$$N_s = \frac{\text{secondary volts}}{\text{volts per turn}} = \frac{330}{0.1}$$

$$= 3300 \text{ turns}$$

Again note that

$$\frac{N_p}{N_s} = 20$$

2 *a)* Figure 5.1 shows the connections of an auto-transformer. If its primary winding has 500 turns and it is connected to a supply of 240 V, what are its secondary voltages if tappings are made at (i) 100 turns, (ii) 250 turns and (iii) 600 turns?

  b) What are the dangers in this type of transformer when connected to a public supply?

Solution

  a) (i) $V_s = V_p \times N_s/N_p$
        $= 240 \times 100/500 = \underline{48 \text{ V}}$
    (ii) $V_s = 240 \times 250/500 = \underline{120 \text{ V}}$
    (iii) $V_s = 240 \times 600/500 = \underline{288 \text{ V}}$

  b) The danger would be if the autotransformer's winding became open circuit as this voltage may appear across the secondary terminals.

3 An autotransformer has a transformation ratio of 10:1. If its primary voltage is 20 V and secondary load current is 20 A (ignoring losses) determine:
  a) the primary current and secondary voltage
  b) the current flowing in the secondary winding.

Solution

  a) Using the transformation ratio:
    $V_p/V_s = I_s/I_p$
    then $V_s = V_p/10 = 200/10 = \underline{20 \text{ V}}$
    and $I_p = I_s/10 = 20/10 = \underline{2 \text{ A}}$

  b) Reference to Figure 5.1 shows that in this section of the transformer the primary current is subtracted from the secondary load current.
    Thus $20 - 2 = \underline{18 \text{ A}}$

4 *a)* Explain what is meant by the 'power factor' of an a.c. circuit.
  b) State two disadvantages of a low power factor.
  c) The rated output of a single-phase transformer is 6 kVA, 240 V. Calculate:
    (i) the full-load current
    (ii) the full-load power at unity power factor
    (iii) the full-load power and reactive voltamperes at 0.8 power factor lagging.

Solution

  c) (i) $I = 25A$
    (ii) $P = 6kW$
    (iii) $P = 4.8kW$
        $Q = 3.6kVA_r$

Instrument transformers

These are transformers used in conjunction with measuring instruments, since in large power systems involving high voltages and large currents it would be difficult and uneconomical to design ordinary instruments for such purposes. There are two types of instrument transformer normally used, namely the *current transformer (c.t.)* and the *potential transformer (p.t.)*. They both have advantages in that not only do they reduce the system's voltage and current to a safe level but they also involve smaller-size cables between themselves and the measuring instruments. This provides a certain degree of isolation from the measuring instruments and allows them to be used in places of convenience such as found in industrial switch rooms or power station control rooms.

The current transformer may either consist of a small number of primary windings or be of the *bar-primary type* as shown in Figure 5.7(b). Here it is

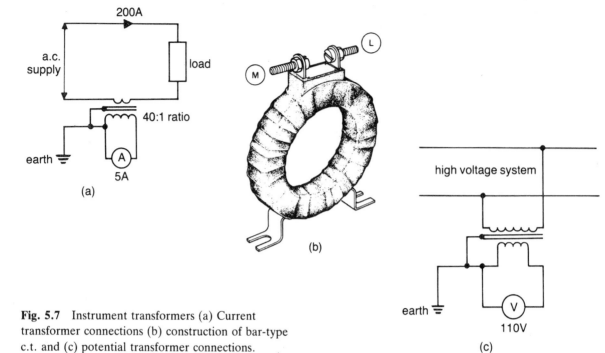

Fig. 5.7 Instrument transformers (a) Current transformer connections (b) construction of bar-type c.t. and (c) potential transformer connections.

normal practice to pass the c.t. over a main *bus-bar* section and connect its secondary terminals marked M and L to the measuring instrument. The arrangement allows the secondary winding to read a proportional amount of the system's large current and it is common practice for the secondary circuit to be standardized at 5 A max. If the main circuit full-load current were 500 A, then a c.t. with a transformation ratio of 100:1 would be chosen.

It is most important to understand that the secondary circuit *must never* be opened while the primary circuit is carrying current. This is because the secondary magnetic flux is being used to stabilize the primary magnetic flux – as was pointed out earlier. If the secondary circuit became open circuit as a result of disconnecting an instrument a large voltage would be induced in the secondary winding. Apart from being dangerous, this situation would cause excessive heating of the c.t. and possibly lead to breakdown of its insulating. To overcome this problem it is normal practice to short out the c.t. first before disconnecting the measuring instrument.

The potential transformer is somewhat larger than the c.t. and is used for reducing the system's voltage.

For safety reasons this is standardized at 110 V. Figure 5.7(c) shows the connection of p.t.

Note: These transformers will be shown in a later chapter.
N.B. See Chapter on instrument connections.

Basic theory of semiconductors

A semiconductor material is one whose electrical conductivity lies somewhere between that of good conductor and that of a good insulator. Two of the most common materials used today are silicon and germanium, with silicon having the better thermal stability.

The electrons in an atom's outer orbit or shell are known as *valence electrons*. Each atom is arranged in a regular pattern with its four valence electrons interchanging with four valence electrons from four neighbouring atoms, as shown in Figure 5.8. The paths of interchange made by the electrons are called *covalent bonds*. At normal room temperature, valence electrons will break away from their covalent bonds and wander throughout the semiconductor crystal. When this happens to a particular atom it

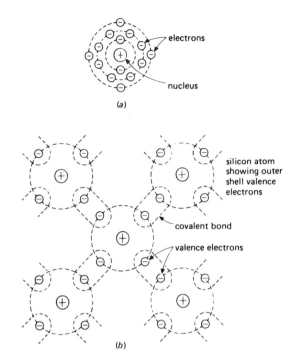

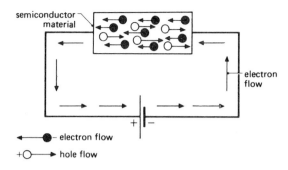

Fig. 5.9 Electron flow and hole flow in a semi-conductor material.

Fig. 5.8 (a) Silicon atom (b) Silicon crystal lattice.

leaves behind an electron vacancy in the form of a *hole*. Since electrons are negatively charged particles the region left behind is often referred to as a *positively charged hole* (remembering that an electron leaving an atom creates a positively charged ion).

When a potential difference is applied across a piece of semiconductor material, free electrons will drift towards the positive potential while holes will give the appearance of drifting towards the negative potential. The number of free electrons and holes will always remain constant. Free electrons will continue to drift around the whole circuit constituting a current flow, but the holes can move only within the semiconductor crystal – see Figure 5.9.

In order to overcome the relatively poor conductivity of the semiconductor material without purposely increasing its temperature (thus making more free electrons and holes available), impurities are added to it in a process known as *doping*. Semiconductors in their pure state are called *intrinsic semiconductors*, but when a controlled amount of an impurity is added they are called *extrinsic*

semiconductors. Two groups of impurity are used, they are:

a) *Pentavalent impurities* – materials such as arsenic, antimony or phosphorous containing *five* valence electrons, and

b) *Trivalent impurities* – materials such as aluminium, gallium or indium containing *three* valence electrons.

When pentavalent impurities are added they *donate* excess electrons to the semiconductor material. In this way they give it a surplus of negative charge carriers – more often called *majority charge carriers*. The material doped in this way, i.e. where the electrons greatly outnumber the holes, is called *n-type extrinsic material*. When trivalent impurities are added, there is a tendency for the impurity atoms to 'rob' electrons from neighbouring semiconductor atoms, so creating an electron deficiency and leaving the material with a surplus of holes. These holes are positive since they belong to fixed positive ions, and in this particular instance they are the majority charge carriers. Material doped in this way is called *p-type extrinsic material*. The lattice structure for both n-type and p-type material is shown in Figure 5.10. In n-type material, the free electron is surplus and not tied to any covalent bond. It will drift about in the crystal and the impurity atom will become a *positively charged donor ion*. In the p-type material, the hole will be filled by any of the neighbouring valence electrons, making the impurity atom a *negatively charged acceptor ion*.

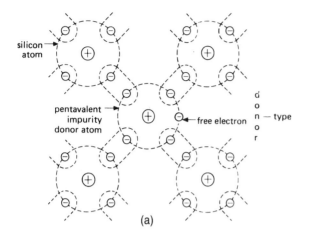

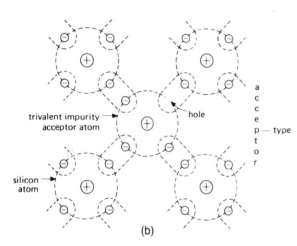

Fig. 5.10 (a) n-type material (b) p-type material.

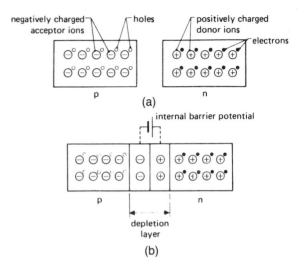

Fig. 5.11 (a) Separate p and n materials (b) pn junction.

Diode

A semiconductor diode is created when both p-type and n-type materials are brought together to form a pn junction. From Figure 5.11 it will be seen that electrons and holes close to the junction, cross over and recombine, leaving a very thin region depleted of both electrons and holes. This region is known as the *depletion layer*. The loss of holes in the p-region near the junction, exposes negatively charged acceptor ions, while the loss of electrons in the n-region near the junction, exposes positively charged donor ions. The congregation of both these unlike charges gives

rise to a *barrier potential* similar to that of a small cell with a p.d. of a few tenths of a volt.

If any external voltage is applied across the pn junction, as shown in Figure 5.12 the battery connections act in opposition to the barrier potential, such that the positive terminal of the battery repels holes and the negative terminal repels electrons. If the battery voltage is sufficiently large it will neutralize the internal barrier potential causing majority charge carriers (holes and electrons) to cross over the junction. The current in the semiconductor diode is due to the hole flow in the p-region and electron flow in the n-region with both holes and electrons flowing at the junction. In this mode of connection, the diode is said to be *forward biased*, i.e. current flows in the forward direction.

If the polarity of the battery is reversed, as shown in Figure 5.13, the holes in the p-region are attracted towards the negative terminal of the battery and the free electrons towards the positive terminal. The result of this is to leave a wide depletion layer in which there are no holes or free electrons – apart from those produced by thermal agitation called *minority carriers*. Thus, in this mode of connection very little current flows around the circuit and the diode is said to be *reverse biased*.

Figure 5.14(c) shows the current–voltage

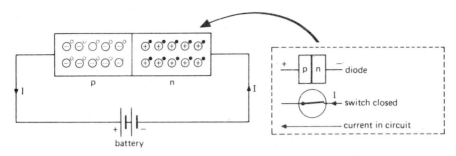

Fig. 5.12 No internal barrier potential as a result of forward bias.

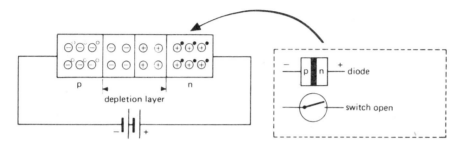

Fig. 5.13 Increased depletion layer as a result of reverse bias.

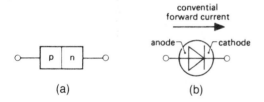

(a)

(b)

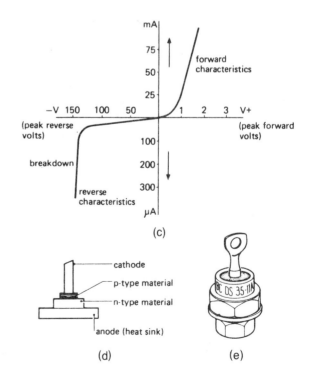

Fig. 5.14 (a) Diode block structure
(b) Symbol
(c) Typical diode characteristics
(d) General structure
(e) Typical appearance of power diode.

characteristics of a typical semiconductor diode. When it is forward biased a small p.d. is applied across the junction to overcome the barrier potential. Because of the diode's low resistance, current through it will increase and so will its temperature. It is important, therefore, that precautions are taken to see that its maximum current rating is not exceeded.

When the diode is reverse biased the reverse current due to minority carriers soon reaches its maximum value. This will occur when the minority carrier rate of flow equals the rate at which they are produced as a result of thermal breakdown. This maximum current or saturation current is often referred to as the diode's *leakage current*. As the voltage in the reverse direction is increased beyond a critical value the diode will break down and become permanently damaged. The maximum safe value of voltage which can be applied before breakdown occurs is called the *peak inverse voltage* (p.i.v.). This value is important where the diode may have to withstand a reverse voltage much higher than the r.m.s. supply voltage. A typical semiconductor diode is shown in Figure 5.14(d) and (e).

Half-wave rectification

Used as a half-wave rectifier the pn junction diode allows current to flow (conventionally) in the forward direction through the load on alternate half cycles of the a.c. supply. It acts as a kind of switch, closing when forward biased and opening when reverse biased. This condition occurs when the anode is made positive with respect to the cathode (the switch being closed) and then opposite when the cathode is made positive with respect to the anode (the switch being open). The current passes through the diode in a series of positive half cycles as shown in Figure 5.15.

Full-wave rectification

The method of rectifying the a.c. supply is widely used. From Figure 5.16 it will be seen that the anodes associated with the two diodes are connected to either side of the secondary winding of the supply

transformer while both cathodes are joined together. The anode voltages are 180° (elect) out of phase with each other. When D1 is positive, D2 is negative and vice-versa. D1 will conduct current to the load resistor on the first half-cycle while D2 blocks or acts as an open switch; then on the second half-cycle D2 will conduct current to the load resistor while D1 blocks. In this way current through the load resistor is unidirectional. Unfortunately, however, the output voltage is of a fluctuating nature. The method used for overcoming this is to add a smoothing or filtering circuit as shown in Figure 5.17.

To avoid the expense of providing an accurately centre-tapped transformer, full-wave rectification may be obtained using a bridge circuit containing four diode elements. This arrangement is shown in Figure 5.18. It will be seen that when point X is positive with respect to point Y, diodes D2 and D4 are forward biased while diodes D1 and D3 are reverse biased. Again, D2 and D4 act like closed switches and D1 and D3 act like open switches, consequently, current flows as indicated by the filled-in-arrows. When point Y becomes positive with respect to point X, the diodes' switching mode becomes reversed, with D1 and D3 forward biased and D2 and D4 reverse biased. The current direction in this case is shown by the open arrows. The output voltage is the same as shown in Figure 5.16.

Today, semiconductor diodes have many applications in microwave and electronic circuitry as well as in switched-mode power supply duty, as commutating diodes in half-controlled circuits, and for high frequency duty in chopper and fast switching circuits. Besides the low- and medium-power diodes of ratings between 2.5 A and 150 A, high-power diodes are also available with mean forward currents up to 2000 A.

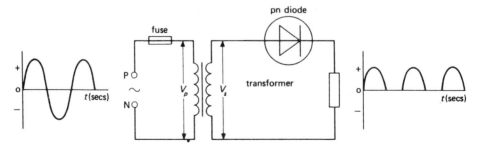

Fig. 5.15   Half-wave rectification using a pn junction diode.

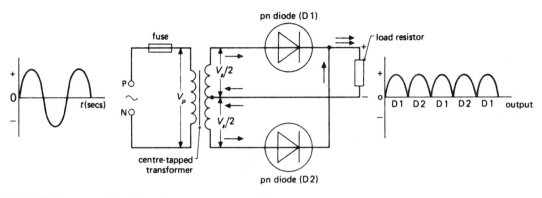

Fig. 5.16 Full-wave rectification using a centre-tapped transformer and two diodes.

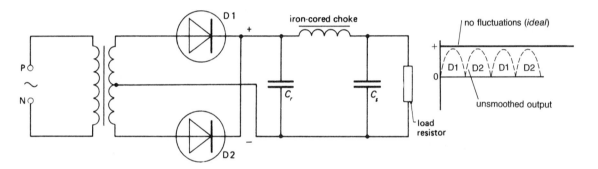

Fig. 5.17 Full-wave rectifier with smoothing circuit:
  C_r is reservoir capacitor,
  C_s is smoothing capacitor.

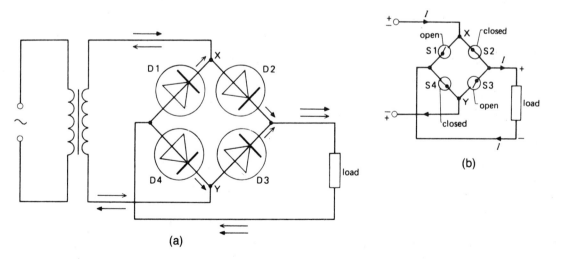

Fig. 5.18 (a) Full-wave bridge rectifier circuit
  (b) Current path when X is positive with respect to Y, using switch analogy.

Zener diode

In order to control or to maintain a constant voltage across the load, a semiconductor device known as a zener diode is sometimes used. It is often referred to as a voltage regulator or stabilizer designed to breakdown at a given voltage. A typical cicuit for the device is shown in Figure 5.19. The current limiting resistor is carefully selected and the voltage drop across it is proportional to the total current taken from the supply. The zener diode is connected in the reverse bias mode.

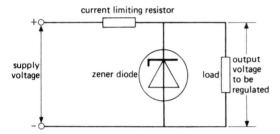

Fig. 5.19 Use of zener diode to give voltage regulation.

If the supply voltage increases then the total current in the circuit will increase. Since the voltage drop across the zener diode will remain constant, or nearly so, the increase in the supply will appear entirely across the limiting resistor. The increased current which flows as a result of the voltage increase will appear in the limiting resistor and the zener diode. Since the voltage across the load remains constant due to the diode, the current in the load will be unaffected by the supply voltage increase.

If the supply voltage decreases, then the total current in the circuit will decrease. Providing that the output voltage does not fall below the critical zener voltage of the diode, the reduced current will cause a corresponding reduction in the voltage drop across the limiting resistor, allowing the load and zener voltage to remain constant. The current in the load is unaffected by the supply voltage decrease and the current reduction takes place only in the series resistor and the zener diode.

Transistor

A junction transistor, or as it is more commonly called a *bipolar transistor*, is obtained when two semiconductor diodes are joined back to back. Depending upon the order in which they are doped, they may be arranged as an *npn-type transistor* or as a *pnp-type transistor*. They are both very similar, with the npn-type mainly conducting electrons and the pnp-type mainly conducting holes. The basic construction details are shown in Figure 5.20, where it will be seen that they provide three connection regions, namely *emitter*, *base* and *collector*. It is not the intention of this chapter to explain their operation fully but from a practical point of view they find their greatest use in radio and amplifier circuitry. Their three modes of operation are shown in Figure 5.21. It will be seen that the input is connected between one pair of terminals and the output taken from another pair. One terminal is always common to both input and output. Each circuit has its own particular characteristics which dictate its application.

Thyristor

There are several types of thyristor, the most common being the reverse blocking triode thyristor called the silicon controlled rectifier (SCR) or simply thyristor. Other members of the thyristor family include the bidirectional diode thyristor (DIAC), bidirectional triode thyristor (TRIAC) and silicon controlled switch (SCS). Here the word thyristor will refer to the SCR. The first syllable of the word 'thyristor' is Greek, meaning a *door*. Since it is a device with door-like characteristics, it can be open or shut at will, preventing or permitting the flow of current in a circuit.

The thyristor, unlike the diode, has a third terminal known as a *gate*. It can be compared with a switch since it performs the same basic function. Forward conduction can occur only when the supply is of the correct polarity, due to the rectifier action. This is initiated by the action of a small pulse of current into the gate supplied as required by a trigger circuit. Conduction can only be stopped when the current flow is interrupted – by opening the circuit

structure   connections

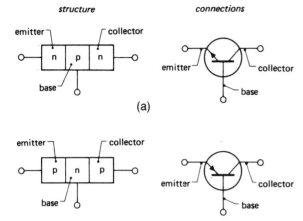

(a)

Fig. 5.20   (a) npn type transistor
(b) pnp type transistor.

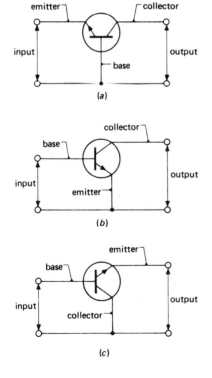

Fig. 5.21   (a) Common base connection
(b) Common emitter connection
(c) Common collector connection.

with S2 or (on a.c) when the current passes through zero (see Figure 5.22).

An alternative way of starting conduction of the thyristor is by applying a bias voltage to the gate with respect to the cathode. However, without the application of the trigger pulse or grid bias no conduction can occur, but once flowing the current is in no way influenced by the gate.

A typical thyristor characteristic is shown in Figure 5.23, and it will be noticed that it can be fired (or triggered) at any desired value of forward voltage by

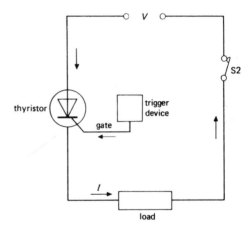

Fig. 5.22   Thyristor circuit with means of supplying gate signal.

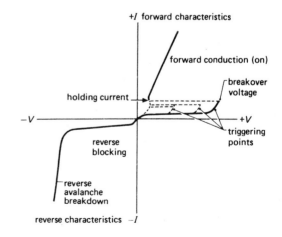

Fig. 5.23   Thyristor characteristics.

the application of a gate signal. Thyristors can be switched from one state to the other in a few micro-seconds, and considerable current gain is available between the gate and the anode of the device: this means that a very small gate power can be used to control a high power in the load circuit. It is not unusual to find thyristors being used for loads ranging from 1 A to 1200 A with peak reverse voltages of up to 4000 V. Further details of the thyristor are given in Figure 5.24 and Figure 5.25 shows two examples of thyristor gate triggering.

Figures 5.26 and 5.27 are two simple examples of how the thyristor is used on d.c. and a.c. supplies. In Figure 5.26 the thyristor is incorporated in a d.c. on–off lamp circuit. The circuit is made to stay on when S1 is pressed and can only be switched off from S2 (opening the circuit as shown or shorting out the thyristor shown dotted). The circuit shown in Figure 5.27 connects the thyristor to the a.c. supply and this gives half-wave power to the lamp. The diode prevents reverse bias being applied to the gate on negative half-cycles. When S1 is closed, the thyristor

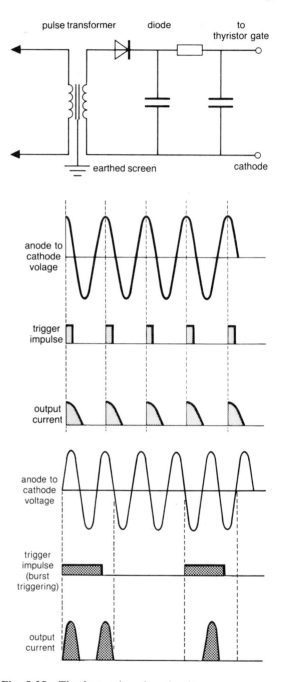

Fig. 5.25 Thyristor triggering circuits.

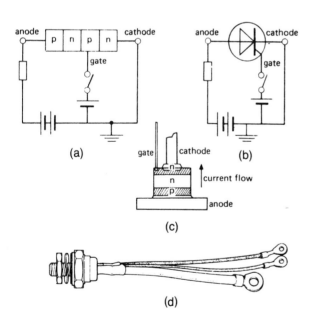

Fig. 5.24 (a) Thyristor block structure and circuitry
(b) Symbol and circuitry
(c) Construction
(d) Typical appearance of power thyristor.

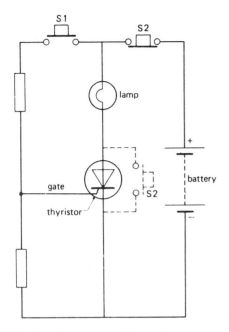

Fig. 5.26 Simple d.c. on-off lamp circuit.

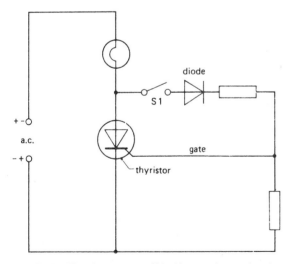

Fig. 5.27 Simple a.c. on-off half-wave lamp circuit.

stays off at the start of each positive half cycle of the supply, but shortly afterwards sufficient voltage is available to trigger the thyristor and the lamp lights up. When the thyristor passes current its anode

voltage falls to practically zero and therefore the gate triggering is removed. The thyristor remains on for the rest of the half-cycle until it is automatically turned off, its anode current then falls to zero. The process is repeated while S1 stays closed.

Triac and diac

The triac has forward and reverse characteristics similar to the forward characteristics of the SCR. The device can block voltages of either polarity and also conduct current in either direction, just like two thyristors in inverse parallel connection, The diac is a triggering device having a high turn-on voltage (about 35 V). It acts as an open switch until the applied voltage reaches the value mentioned and then it triggers, sending a pulse across the load. The diac will turn off if the current falls below its minimum holding value.

Figure 5.28 shows how both are connected in a basic lamp dimmer circuit. Briefly, it will be seen that both *R* and *C* are connected across the a.c. supply, acting as a combined variable voltage arrangement. Adjustment of *R* alters the p.d. across itself and across *C*. The components provide the network with a variable voltage and a variable phase shift.

The diac is connected between *R* and *C* and when the p.d. across *C* rises it turns on the diac which pulses the triac gate and allows it to conduct. Because the triac is connected to the a.c. supply, turn-off occurs automatically as the a.c. cycle passes

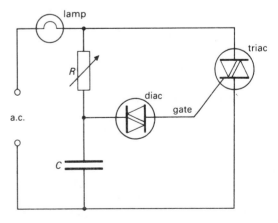

Fig. 5.28 Basic lamp dimmer circuit.

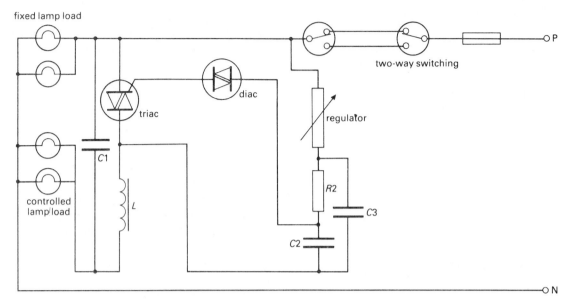

Fig. 5.29 Two-way dimmer switch control.

through zero. Where R is set to its minimum, the triac fires soon after the start of each half-cycle and supplies full power to the lamp. Where R is set to its maximum, the firing of the triac is delayed because of the low p.d. across C and phase shift, consequently, minimum power is supplied to the lamp. An improved circuit is shown in Figure 5.29, and is typical of modern lamp dimmers.

Three-phase rectifiers

Knowledge of three-phase rectifiers is not required by Part II students and only a brief mention will be made. Figure 5.30(a) shows a method of obtaining three-phase half-wave rectification using a rectifier in each phase. They are seen star connected supplying a single-phase load. The output waveform is shown in Figure 5.30(b). When the red phase is positive with respect to the yellow and blue phases, rectifier 1 will conduct. Rectifier 2 will conduct when the yellow phase is positive with respect to the other two phases, and then rectifier 3 will conduct. It will be noticed that the rectified waveform never reaches zero as was the case for the single-phase, half-wave rectifier. However, comparison between the two will show a smoother output is produced using three-phase.

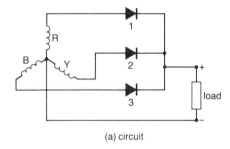

(a) circuit

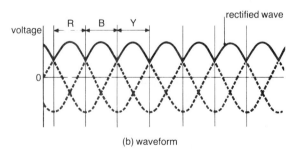

(b) waveform

Fig. 5.30 Three-phase half-wave rectifier.

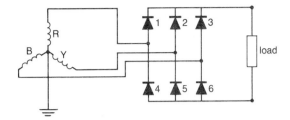

Fig. 5.31 Three-phase bridge rectifier.

In order to increase the smoothness of this output waveform, six or more rectifiers are used. This can be achieved by a centre-tap arrangement or bridge arrangement. Figure 5.31 shows a typical circuit of a three-phase bridge rectifier and the whole of the voltage across each phase winding is utilized at any instant in time. This method not only gives a smoother output but is also used to provide a higher output voltage. The rectification sequence is that when the red phase is positive with respect to the other two phases, rectifier 1 will conduct and current will flow through the load and return to the transformer via rectifiers 5 and 6. When rectifier 2 conducts, current flows through the load and returns via rectifiers 4 and 6. And when rectifier 3 conducts, current flows through the load and returns via rectifiers 4 and 5. This process is repeated and whilst the waveform may still show some ripple, it can be greatly reduced using some form of capacitor filter.

Exercise 5

1 Draw the winding arrangement for a single-phase, double-wound, shell-type transformer assuming sandwich windings. Label all parts of the transformer.

2 *a)* A single-phase, double-wound, core-type transformer is designed for a primary voltage of 240 V and a secondary voltage of 12 V. What is the ratio of the transformer?

   b) If the primary winding has 600 turns, how many turns are there on the secondary side?

   c) How many volts per turn on the secondary winding?

   d) How many turns on the secondary for a voltage tapping of 4 V?

3 Draw a circuit diagram of the transformer in Question 2 above and label its windings for the primary and secondary voltages.

4 *a)* Explain with the aid of a diagram how cooling occurs in an oil-immersed transformer.

   b) List *two* possible faults which a transformer might develop.

5 Make a circuit diagram of a three-phase supply transformer showing a delta-connected primary at 11 kV and star-connected secondary at 415 V/240 V. Assume the secondary is to feed a 3-phase, 4-wire distribution system using TN-S earthing.

6 Figure 5.32 shows a thyristor speed control circuit for an electric motor. Identify and state the function of components

   a) T_1

   b) R_1

   c) R_2

   d) D_1

   e) D_2

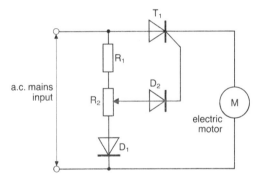

Fig. 5.32 Thyristor control of a motor.

CGLI/II/86

7 Figure 5.33 shows a simple circuit for providing a controlled d.c. supply to a resistive load.

   a) Name the device marked X.

   b) Draw the device marked X and label the terminals.

   c) Sketch three cycles of the a.c. voltage waveform.

   d) On the voltage waveform drawn in (c) superimpose the current waveform over three cycles of the supply voltage waveform,

if the device is arranged to trigger at 45°
from zero voltage.

CGLI//II/86

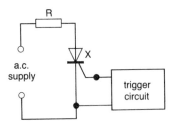

Fig. 5.33

8 *a)* Explain what happens when forward and
         reverse bias is applied to a pn junction
         diode.
   b) What is meant by the term *peak forward
         voltage?*
9 *a)* Explain the purpose of a *heat sink*.
   b) What is the purpose of a *zener diode?*
10 *a)* Draw circuit diagrams to show how direct
         current may be obtained to supply a
         resisitive load from an a.c. supply, using
         (i)   one diode
         (ii)   two diodes to obtain full-wave
                rectification
         (iii) four diodes connected as a bridge
                rectifier.
         On the diagrams indicate the a.c. input, d.c.
         output and polarity.
    b) Sketch the waveform of the output current
          over two cycles for
          (i)   the one diode circuit drawn in (a) (i)
          (ii) the four diode-circuit drawn in (a) (iii).
    c) State *two* operating conditions which can
          destroy the electrical properties of a
          semiconductor diode.

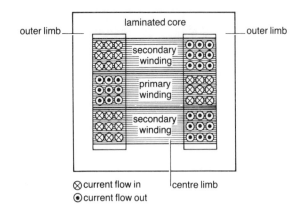

Fig. 5.34

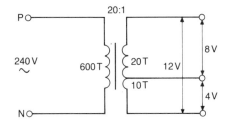

Fig. 5.35

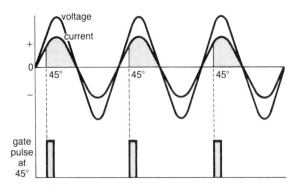

Fig. 5.36

CHAPTER SIX

Heating and lighting calculations

After reading this chapter you will be able to:

1   State the ways heat is transmitted by conduction, convection and radiation.

2   Describe common methods of space heating and water heating and know their advantages and disadvantages.

3   Know methods of heat control through simmerstats and thermostats.

4   Perform calculations associated with heat energy, electrical energy and efficiency of heating systems.

5   State the laws of illumination.

6   Perform calculations on the point-by-point method of illumination using the inverse square law and cosine law.

Heating effects of current

Heat is a form of energy which can be obtained in various ways. This part of the chapter is concerned with the way it is produced as a result of passing current through a resistive element.

It is a well-established fact that heat transference takes place in one or more of the following ways: *conduction, convection* and *radiation*.

Conduction is the transfer of heat through a material from molecule to molecule without any appreciable movement of the molecules themselves – excepting their rapid vibration. Heat transference is said to 'flow' or be *conducted* through the material from a place of higher temperature to a place of lower temperature. Most metals are usually good conductors of heat as well as being good electrical conductors, while air, wood, glass, pvc, rubber, etc. are poor heat conductors as well as poor electrical conductors. There are many examples in the electrical industry where heat conduction is utilized, such as in soldering irons, block storage heaters and water-heating elements.

Convection is the transfer of heat through liquids or gases. The molecules are loosely bound together and those which come in contact with the source of heat become hotter (i.e. less dense) and expand or rise and their place is taken by colder or cooler

molecules. This sets up convection currents which circulate within the transfer medium. Convector heaters, immersion heaters and oil-filled transformers all make use of this method.

Radiation is the transfer of heat in the form of rays or particles and this occurs in straight lines from some source. The infra-red radiation transferred to earth from the sun is a typical example. Heat rays are cold and when this energy hits an object, some of it may be absorbed and the object becomes hotter. For example, dull black surfaces are good absorbers of heat as well as being good radiators. This is why cooling fins on the back of refrigerators are painted black so that they will radiate more heat. In contrast, shiny, bright surfaces are poor absorbers of heat since they reflect heat away. In hot countries, houses are painted white to reduce heat absorption from the sun. There are numerous types of electric fire which have reflecting surfaces and among these will of course be found infra-red heaters. Figure 6.1 gives an idea how a radiant heater transmits its rays in straight lines and shows how some objects partially absorb and partially reflect the rays. Warm surfaces that become heated either conduct or convect the heat to colder objects.

Some electrical equipment may be designed specifically with more than one method of heat

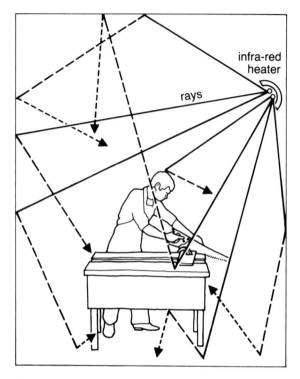

Fig. 6.1 Radiation from infra-red heater.

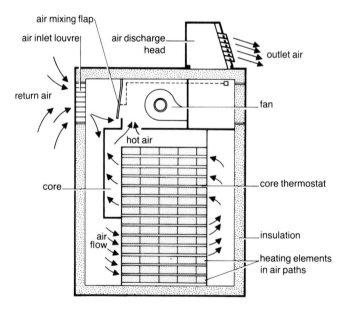

Fig. 6.2 Electricaire unit.

transfer in operation, such as a fan-assisted block storage radiator or an Electricaire unit as shown in Figure 6.2. Here it will be seen that the heating elements conduct heat through the core and then the heat rises by convection, ready to be blown out of the unit by a fan as useful hot air. There are many other forms of electric heating such as ceramic cooking hobbs and induction hobbs and also microwave ovens which use electromagnetic waves at high frequency. In practice, it is possible to divide domestic heating into two categories, that required for *space heating* and that required for *water heating*.

Space heating is normally provided by: (i) *direct-acting heaters* such as convector heaters, wall/ceiling heaters, panel heaters, oil-filled radiators and fan heaters, and (ii) *thermal-storage heaters* such as storage radiators, Electricaire units and under-floor heating systems.

Water heaters can be provided by *open outlet heaters, cistern-type heaters* and *cistern-fed heaters*. These systems have been described in *Electrical Installation Technology 3 – Advance Work* by the same author and be dealt with only briefly in this chapter.

Space heating

Direct-acting heaters are ideal for premises which have irregular or infrequent use where instant heat can be switched on and directed to the area most needed. Thought should be given to the type(s) which can be switched off automatically when no longer needed. Normally, such heaters provide convenience and immediate response and use electricity at the standard rate. As an approximate guide, a room of 15 m^2 floor area and 2.75 m in height with normal windows and cavity walls will require at least 2 kW of heating to maintain a temperature difference of 20°C. Note: This is for a room having a total outside wall length of 5.0 m.

Thermal storage heaters are beneficial to those premises having long periods of occupation and they take advantage of the cheap night-rate electricity referred to as *Economy 7*. Modern systems are very efficient and compact and are provided with cost-saving devices such as charge controllers and thermostats to give the right amount of heat according to their installation environment. Such

systems operate by storing up heat during the off-peak period and then releasing it as and when required throughout the day period. Using the same data as above for direct-acting heaters and based on the cheap seven-hour tariff, at least 5 kW of night-storage heating is required.

Water heating

Open outlet type heaters are those which are designed to connect directly to the cold water supply where the hot water is needed. They can be installed oversink or undersink and they usually store small quantities of water. The *instantaneous water heater* is typical of this type of single point heater and is now electronically designed to give reduced running costs and save water. It is suitable for hand wash-basins and showers (see Figure 6.3).

Cistern-type water heaters are used where there is no separate water feed tank. They are in fact connected to the main water supply and since they are gravity fed systems they are most suitable for supplying several taps and hand wash-basins (see Figure 6.4).

Cistern-fed water heaters are multipurpose systems supplied from a separate cistern mounted at high level. The system feeds a storage tank which can be provided with immersion heaters utilizing cheap 'off-peak' electricity. Their main advantage is the large capacity of heated water that can be stored (see Figure 6.5).

The storing of much larger quantities of heated water might involve the use of *heat pumps* or *electrode boilers*. In the former system, heat is extracted from the air in an evaporator unit before it is passed into a compressor which pumps the water from a low temperature to a high temperature into a storage tank ready for use. In the latter system, heat is generated by the passage of current through the water and control is effected by raising or lowering immersed electrodes (see Figure 6.6).

Temperature

Temperature is a measure of how *hot* or *cold* an object is and must not be confused with heat. The instrument for measuring temperature is called a

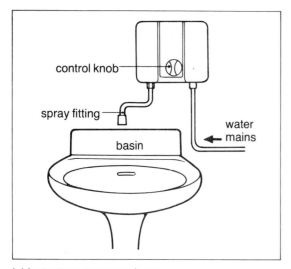

(a) instantaneous water heater

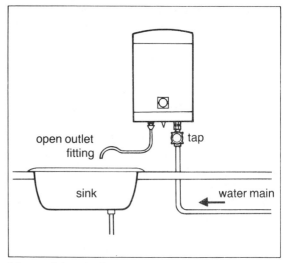

(c) oversink water heater

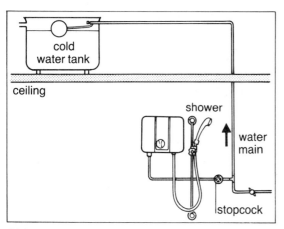

(b) instantaneous shower

Fig. 6.3   Open outlet type water heaters.

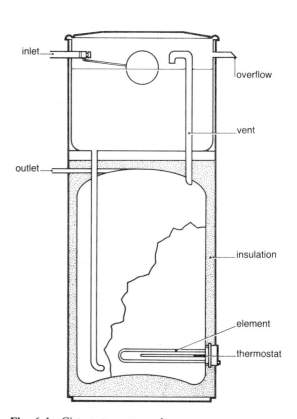

Fig. 6.4   Cistern-type water heater.

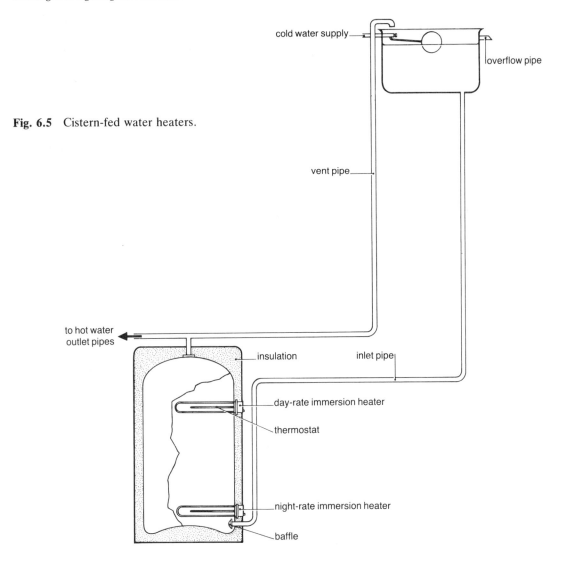

Fig. 6.5 Cistern-fed water heaters.

thermometer and whilst there are many different types such as gas thermometers, pyrometers and thermocouples, the one shown in Figure 6.7 is typical of the type that is likely to be found in a school or college laboratory. It will be seen that mercury is used as the expansion liquid since it expands evenly and is also a good conductor of heat.

In practice, one comes across two temperature scales, namely, *Celsius* and *Fahrenheit*. The former scale is often called the Centigrade scale and is the one used in electrical work. It ranges between two

fixed points, 0°C to 100°C (i.e. the freezing point or temperature of pure melting ice and the temperature of steam while over pure boiling water). As objects or substances get colder and colder, such as when gases condense into liquids and liquids freeze into solids, then the *Kelvin* thermodynamic temperature scale is chosen. This begins at absolute zero (−273°C) and increases like the Celsius scale. It is not marked in degrees, so that 0°C is equivalent to 273 K and 100°C is equivalent to 373 K.

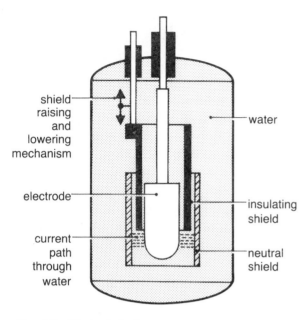

shield raising and lowering mechanism

electrode

current path through water

water

insulating shield

neutral shield

Fig. 6.6 Electrode boiler.

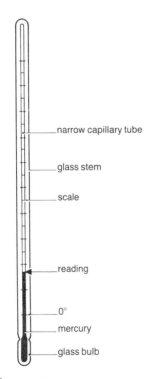

narrow capillary tube

glass stem

scale

reading

0°

mercury

glass bulb

Fig. 6.7 Thermometer.

Specific heat capacity

The specific heat capacity (*c*) of a substance is the amount of heat in joules (J) that is needed to raise the temperature of one kilogram (1 kg) of the substance by one degree Celsius (1°C). Thus *c* = J/kg°C *Note*: Students may sometimes see questions which express specific heat capacity in terms of J/kg K or even kJ/kg K; for example, water could be expressed as 4.2 kJ/kg K. Also the use of the word 'specific' implies that the substance is divided by its mass and the term 'specific heat capacity' really means the heat capacity per unit mass. Out of interest, the term *specific gravity* mentioned in Chapter 2 dealing with cells was used to express the ratio of the density of an electrolyte to that of water and it should really be called *relative density*.

It is found from experiments that the amount of heat needed to raise 1 kg of water 1°C is 4200 J. Copper requires 380 J and mercury requires only 140 J. These values are their specific heat capacities and water would be expressed as 4200 J/kg°C, copper 380 J/kg°C and mercury 140 J/kg°C. They tell us that water, for example, needs more energy to heat it up than copper and mercury and also that it stores more energy when it is hot compared with them. This is why it is such a good liquid to have in central heating systems. From the above it follows that the amount of heat needed to cause a given temperature rise is expressed as:

Heat energy (Q) = specific heat capacity × mass × change in temperature

$$\text{or } Q = c \times m \times (t_2 - t_1) \text{ joules} \qquad (6.1)$$

where t_1 and t_2 are respectively the initial and final temperatures in degrees Celsius.

Example 1

An electric heater is fitted to a copper kettle of mass 0.5 kg containing 1.5 litres of water. Determine the amount of heat needed to raise the temperature of the water from 20°C to boiling point.

Solution

The heat required for the copper kettle

$$Q = cm(t_2 - t_1) \text{ joules}$$
$$= 380 \times 0.5 \times 80 = \underline{15\ 200\ J}$$

The heat required for the water

$$Q = 4200 \times 1.5 \times 80 = \underline{504\ 000\ J}$$

The total heat required is 15.2 kJ + 504 kJ
$$= \underline{519.2\ kJ}$$

In this example, if the kettle had an element rated at 2.5 kW and it was 95% (0.95 per unit) efficient, the time taken for it to boil the water would be calculated as follows:

Since efficiency (η) = output/input
where output is the heat energy (Q) required in joules
and input is the electrical energy (W) required in joules, then input = output/efficiency
Expressed as kilowatt-hours since 1 kWh = 3.6 MJ

then $\quad\quad\quad W = 3 \text{ kW} \times h$
and $\quad\quad\quad Q = 0.5192/3.6 = 0.1442 \text{ kWh}$
Thus $\quad\quad 3 \times h = 0.1442/0.95$
therefore $\quad\quad h = 0.1442/\ (0.95 \times 3)$
$$= \underline{0.05 \text{ hours}} \text{ (3 mins)}$$

It should be noted that heat is lost in the kettle by conduction, convection and evaporation as the water turns into a vapour without actually reaching boiling point.

Questions on heat do not always include the calculation of the container's heat and as seen from the above example, it is the water or liquid which requires the most heat.

Consider some further examples, remembering that 1 litre of water is equivalent to 1 kilogram.

Example 2

A tank containing 1300 litres of water is heated from an initial temperature of 15°C to 55°C. If the electrical supply is by means of five 3 kW heaters for a period of ten hours, determine the efficiency of the system if the specific heat capacity of water is 4200 J/kg°C.

Solution

Heat energy required $\quad Q = mc(t_2 - t_1)$
$$= 1\ 300 \times 4200 \times 40$$
$$= 218.4 \text{ MJ} = 60.7 \text{ kWh}$$
Electrical energy required $W = Pt = 15 \times 10$
$$= 150 \text{ kWh}$$
Efficiency $\quad\quad\quad (\eta) = Q/W = 60.77/150$

$$= \underline{0.4 \text{ per unit}} (40\%)$$

Example 3

An electrically heated instantaneous shower supplies 5 litres of water per minute at a temperature of 30°C. If the initial temperature of the water is 12°C, what is the rating of the shower heater, neglecting losses?

Solution

$$Q = mc(t_2 - t_1)$$
$$= 5 \times 4200 \times 18$$
$$= 378\ 000 \text{ J}$$
$$= 0.378 \text{ MJ/min}$$
Since there are no losses $\quad W = Q = 0.378 \text{ MJ/min}$
Over a period of one hour $W = 22.68 \text{ MJ}$
Since $\quad\quad\quad\quad\quad\quad W = Pt$
and $\quad\quad\quad\quad\quad P = W/t = 22.68/3.6$
$$= \underline{6.3 \text{ kW}}$$

Example 4

Compare the cost of the above shower heater with a bath which uses 135 litres of water at the same temperature. The bath's water heating system is 90% efficient and the shower heater uses 15 litres of water in 3 minutes. Assume the cost of heating the water using the shower is at 5.16 p per unit and the bath system is at the Economy 7 rate of 1.9 p per unit.

Solution

For the shower unit:
Since 3 min = 0.05 h, it uses 6.3 × 0.05
= 0.315 kWh
The cost at 5.16 p/unit is 5.16 × 0.315 = <u>1.6 p.</u>
For the bath system:

Heat required $Q = mc(t_2 - t_1)$ $= 135 \times 4200 \times 18$
$= 10.2$ MJ
As it is 90% efficient, $W = Q/\eta = 10.2/0.9$
$= 11.34$ MJ
Since 1 kWh $= 3.6$ MJ, then $W = 11.34/3.6$
$= 3.15$ kWh
The cost at 1.9 p/unit is $1.9 \times 3.15 = \underline{6 \text{ p.}}$
This example shows that despite the use of cheap night electricity, it is more costly to have a bath than to have a shower.

Example 5

Determine the resistance of a heating element required to raise the temperature of 5 kg of water from 15°C to 60°C in 30 minutes. Assume the water-heater system has 10% heat losses and the supply is 240 V.

Solution

Heat energy required $Q = mc(t_2 - t_1)$
$= 5 \times 4.2 \times 45$
$= 945$ kJ $= 0.945$ MJ
Since 3.6 MJ $= 1$ kWh $Q = 0.945/3.6$
$= 0.2625$ kWh
Electrical energy required $W = Pt = P \times 0.5$ kWh

Since the system is 90% efficient then
$$W = Q/\eta$$
$$(P \times 0.5) = 0.2625/0.9$$
therefore $\quad P = 0.2625/0.45$
$= 0.583$ kW
Since $P = V^2/R$ then $R = V^2/P = 240^2/583$
$= \underline{99 \ \Omega}$

Heating controls

It is normal practice to provide a whole heating system with complete and separate isolation, automatic disconnection and earth leakage protection together with equipotential bonding.

Local and automatic control will initally be through time switches and room thermostats built into equipment. Often room thermostats are the preferred choice in terms of general comfort in space heating systems, whereas individual control for a cooking appliance, a kettle or an immersion heater will provide the effective use of electricity to facilitate operation.

Figure 6.8 shows a circuit diagram of an *energy regulator* which is used to vary the energy input into a radiant ring used on a cooker. It does not provide temperature control. It will be seen that the bimetallic strip carries a heater winding which when energised will open contacts of a snap-action switch. The supply to the radiant heater will be cut off and the bimetallic strip will then cool down and cause the contacts to close again. The cycle is repeated at short intervals to give simmering conditions.

Thermostats are basically temperature-operated switches and need to be suited to the heating system they control. Their purpose apart from controlling temperature is to reduce energy consumption. Figure 6.9 is a circuit diagram of a thermostat having an automatic set-back control for night-time use. This device allows the heating system to be maintained at a lower temperature level (about 5°C below the normal set point) and is an ideal arrangement for premises which are left unoccupied. During cold spells it allows the premises to be quickly heated for day-time occupation.

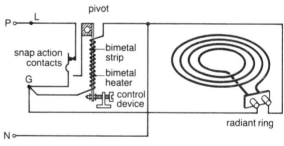

(a) circuit diagram

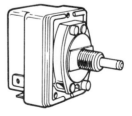

(b) control device

Fig. 6.8   Control of energy.

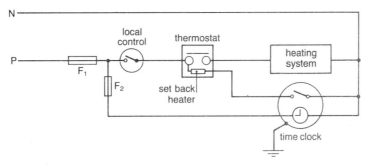

Fig. 6.9 Automatic 'set-back' control circuit for a heating system.

Figure 6.10 shows an immersion heater thermostat which operates on expansion and contraction of two metals with different coefficients of expansion. The diagram is that of a *rod-type* thermostat which comprises a long brass tube and a non-expanding steel rod. Both metals are joined together at the opposite end to the switch and the heating action expands the outer brass tube and releases the non-expanding invar rod. This causes the circuit to open at its correct setting. Inside the switch contacts is incorporated a small permanent magnet which provides a snap action to stop contact wear.

Lighting calculations

Light is a form of energy. The light obtained from the sun or from an incandescent lamp is usually considered to be 'white' light, but in actual fact it is composed of all the colours of the rainbow. The break-up of white light into its component colours is easily demonstrated by passing a beam of light through a glass prism, as shown in Figure 6.11. It will be seen that the red colour is refracted the least and the violet colour refracted the most. This is because light is a *wave* phenomenon and the various colours have different frequencies. The higher the frequency the shorter the wavelength; therefore, out of all the colours making up the visible light spectrum, the violet colour has the shortest wavelength. The relationship between wavelength (λ) and frequency (f) is given by the following expression, remembering that all light waves travel through space at the same speed or velocity (v).

$$\lambda = \frac{v}{f}$$

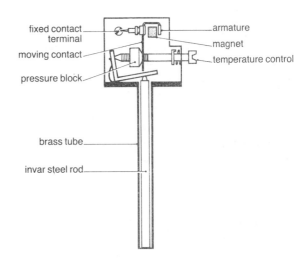

Fig. 6.10 Immersion heater thermostat.

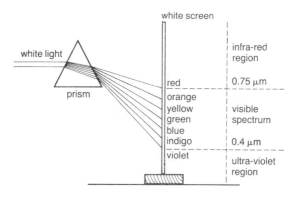

Fig. 6.11 The light spectrum.

where λ is the Greek letter lambda standing for
     wavelength (m)
    f is the frequency (Hz)
    v is the velocity of light travelling
      at 3×10^8 m/s

In order to find the wavelength of, say, the violet
colour, we know its frequency to be 7.5×10^{14} Hz,
therefore

$$\lambda = \frac{3 \times 10^8}{7.5 \times 10^{14}}$$

$$= 0.000\ 000\ 4\ \text{m}$$

$$= 0.4\ \mu\text{m}$$

Wavelength is often expressed in microns (where
1 micron = 10^{-6} m) or ångström units (Å) (where
1 Å = 10^{-10} m). Thus, the violet wavelength could be
written as 0.4 microns or 4 000 Å.

It will be seen from Figure 6.12 that the visible
light spectrum occupies only a small portion of the
electromagnetic radiant energy spectrum. The human
eye will not respond to radiation unless the
wavelength is between 0.4 and 0.75 μm. In Figure
6.13 it will be seen that its sensitivity is highest in the
middle of the response curve, i.e. in the green and
yellow waveband regions. Outside these waveband
regions the curve tails off into the *infra-red* and
ultra-violet zones. Both of these are present in natural
daylight, as well as being emitted by artificial light
sources. Incandescent filament lamps mostly produce
infra-red radiation giving the effect of heat, whereas
ultra-violet radiation is used to excite fluorescent
powders in fluorescent discharge lamps.

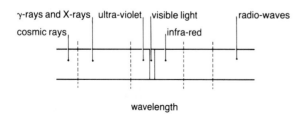

Fig. 6.12 The radiant energy spectrum.

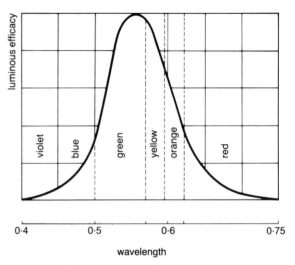

Fig. 6.13 Response curve of the human eye.

Laws of illumination

The following are some of the terms and units used
in lighting calculations.

Term	Unit
luminous intensity (*I*)	candela (cd)
luminous flux (*F*)	lumen (lm)
illuminance (*E*)	lux (lx)
luminance (*L*)	nit (nt)

Meanings

Luminous intensity– the illuminating power of a light
source
Luminous flux – the flow of light measured in lumens
Illuminance – the measure of light falling on a surface
Luminance – the measured brightness of a surface
Figure 6.14 is an illustration of the lumen or rate of
flow of light. A point source is set at the centre of a
sphere of 1 m radius. If the light source has an
intensity of 1 candela, then 1 lumen will be emitted
on 1 square metre of surface, giving an illuminance of
1 lux. Because the surface area of a sphere is $4\pi r^2$
and *r* is 1 m, then 4π lumens are emitted by
1 candela.

There are two important laws used in illumination
calculations, namely the *inverse square law* and the
cosine law. The first law considers that fact that the

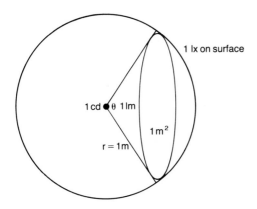

Fig. 6.14 Relationships of light units in a solid angle.

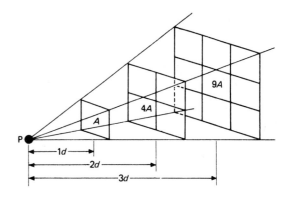

Fig. 6.15 Inverse square law.

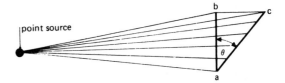

Fig. 6.16 Cosine law.

illumination received on a surface due to a point source is inversely proportional to the square of its perpendicular distance from the source. The second law considers the case where the surface to be illuminated is not perpendicular to the direction of the light source: the light makes an angle with the normal to the illuminated surface. Figure 6.15 and 6.16 illustrate both the laws.

Inverse square law calculations

The inverse square law is given by the expression:

$$E = \frac{I}{d^2} \text{ lux} \qquad (6.2)$$

where d is the distance in metres

Example 1

The illuminance on a surface directly below a point source is 400 lux. If the distance between the light source and the surface is 2 m, what is the intensity of the light source?

From equation (6.2).

$$I = E \times d^2$$
$$= 400 \times 4$$
$$= \underline{1600 \text{ cd}}$$

In this example it is assumed that the point source is a filament lamp of some description, not a fluorescent

tube of some particular length. The inverse square law is not applicable to linear light sources.

Example 2

An incandescent lamp having a luminous intensity of 100 cd in all directions gives an illuminance of 40 lux at the surface of a bench vertically below the lamp.
a) What distance is the lamp above the bench?
b) What illuminance would be received at the bench if the lamp were lowered by 0.58 m?

a) Since $\quad E = \dfrac{I}{d^2}$

then $\qquad d = \sqrt{(I/E)}$
$$= \sqrt{(100/40)}$$
$$= \underline{1.58 \text{ m}}$$

b) $\quad d = 1$ m

therefore $E = \dfrac{100}{1} = \underline{100 \text{ lux}}$

Cosine law calculations

This method considers the illumination at any given point where the illuminance may be the result of one or more lamps or reflection from surrounding walls and/or ceilings.

The cosine law is given by the expression:

$$E = \frac{I}{d^2} \cos \theta \text{ lux} \tag{6.3}$$

Since d is difficult to measure, the height (h) directly below the lamp is used in the calculation.

Since $\cos \theta = \dfrac{h}{d}$ and $d = \dfrac{h}{\cos \theta}$ (see Figure 6.17)

then equation (6.3) becomes

$$E = I \cos \theta / (h/\cos \theta)^2$$

$$E = I \cos^3 \theta / h^2$$

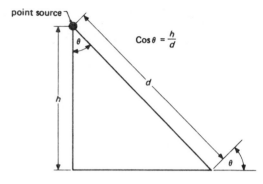

Fig. 6.17 Cosine law geometry.

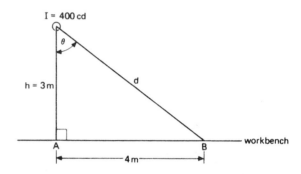

Fig. 6.18

Example 3

An incandescent lamp is suspended 3 m above a level workbench and is fitted with a reflector such that the luminous intensity in all directions below the horizontal is 400 cd. Calculate the illuminance at a point A on the surface of the bench immediately below the lamp and also at a point B, 4 m away from point A (Figure 6.18).

The geometry of the problem is that of a 3-4-5 right-angled triangle, and from equation (6.2) the illuminance at A is given by

$$E = \frac{I}{d^2}$$

$$= \frac{400}{9}$$

$$= \underline{44.4 \text{ lx}}$$

The illuminance at B is given by the expression (6.3)

$$E = \frac{I}{d^2} \cos \theta \text{ lux}$$

therefore $E = \dfrac{400}{25} \times 0.6$

$$= 9.6 \text{ lx}$$

Alternatively $E = \dfrac{I}{h^2} \cos^3 \theta \text{ lux}$

$$= \frac{400}{9} \times 0.6^3$$

$$= \underline{9.6 \text{ lx}}$$

Note: It should be realized that this alternative approach already considers I/h^2 calculation and consequently saves time.

Example 4

In the above example, assume another incandescent lamp of 400 cd is installed directly over point B at a height of 3 m. What would be the illuminance at point A and point B as a result of both lamps? (Figure 6.19).

In this example the illuminance is the addition of the two light sources, since both points share

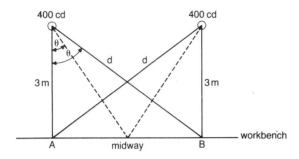

Fig. 6.19 Common intensities.

common heights and common lamp intensities. In this case, point A will receive the same amount of light as point B since the lamps have common intensities and are at common heights. The light from each source is additive, i.e.

illuminance at point A = 44.4 + 9.6 = $\underline{54\ lx}$
and illuminance at point B = $\underline{54\ lx}$

Example 5

With reference to Figure 6.19 determine the illuminance midway between point A and point B on the workbench.

Here, d is approximately 3.6 m and cos θ − 0.83

since $E = \dfrac{I}{d^2} \cos θ$

therefore $E = \dfrac{400}{3.6^2} \times 0.83$

$\qquad = 25.7\ lx$

Since there are two lamps shining on this spot, then the illuminance = $\underline{51.4\ lx}$

Exercise 6

1 Draw a neatly labelled circuit diagram of a domestic consumer's intake position having provision for a 24-hour supply and Economy 7 night supply. The diagram should include:
   a) Electricity Board's two-rate meter, time clock and contractor.
   b) Separate consumer units for the conditions stated.

2 Draw a neatly labelled diagram of a block-storage radiator and explain how heat is transmitted to a room.

3 A water heater is rated at 1 kW/240 V and holds 20 litres of water. Determine the time it takes to raise the temperature from 20°C to 100°C if the efficiency is 80% and the specific heat capacity of water is 4.2 kJ/kg°C.

4 a) Explain how heat is transmitted by convection.
   b) Make a sketch of *two* of the following types of convector heater, briefly describing their operation.
      (i)   oil-filled radiator
      (ii) panel heater
      (iii) tubular heater.

5 Determine the percentage efficiency of a water heater connected to a supply of 240 V and taking a current of 2.6 A. The heater holds 7.2 kg of water and takes 50 min to raise the temperature from 16°C to 76°C.

6 With reference to Figure 6.20, determine the illuminance at B and C.

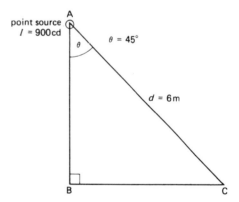

Fig. 6.20 Application of laws of illumination.

7 Two incandescent filament lamps are arranged on a photometer bench as shown in Figure 6.21. The movable photometer head is fitted with a double-sided white matt screen and the side opposite the 60 cd lamp receives an illuminance of 26 lx while the other side receives 67 lx. Find:
   a) the distance each lamp is from the screen
   b) the illuminance on each side of the screen

when it is placed halfway between the lamps.

8 A number of 400 W mercury discharge lamps are used to illuminate a car park.

   a)   (i)   Why is this type of lamp more suitable for outside lighting than a 500 W metal filament lamp?

        (ii)   State *one* major operational disadvantage of this type of lamp.

        (iii)   Why is the total power of *each* luminaire greater than 400 W?

   b)   One lamp was tested, as shown in Figure 6.22. With the lamp operating at full brilliancy and the switch S open, A1 reads 3.5 A and A2 reads zero. With the switch S closed A2 reads 1.6 A. The power in each case is 440 W. Draw a phasor diagram to a scale of 1 A = 2 cm and from it find the current in A1 with switch S closed.

                                 CGLI/II/82

9 *a)*   Figure 6.23 shows the connections of low-pressure sodium vapour discharge lamp (SOX). When it is at full brilliance and with switch S open, the ammeter A1 reads 1.2 A, while the wattmeter reads 68 W. If switch S is closed, A1 reads 0.35 A, the wattmeter remains unchanged and ammeter A2 reads 1 A. Construct a phasor diagram of the circuit under both open and closed conditions of the switch.

   b)   If the lamp emits 5500 lumens and the ballast has losses of 13 W, calculate the efficacy of the lamp only.

   c)   State the reason for the resistor across the capacitor.

   d)   Give two safety precautions to be observed when disposing of sodium lamps.

10 Figure 6.24 shows the connections of a SON discharge lamp. With switch S open, ammeter A_1 reads 5 A and W reads 420 W. With S closed A_2 reads 2.2 A and A_1 reads 3 A, W reads 420 W again.

   a)   Draw a phasor diagram of the circuit using 1 A = 2 cm.

   b)   If the lamp emits 44 000 lm and losses in the circuit are 20 W, what is the efficacy of the lamp?

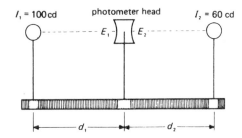

Fig. 6.21 Photometer bench.

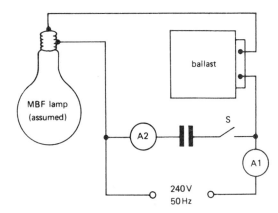

Fig. 6.22 MBF discharge lamp circuit.

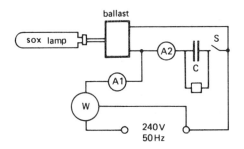

Fig. 6.23 SOX discharge lamp circuit.

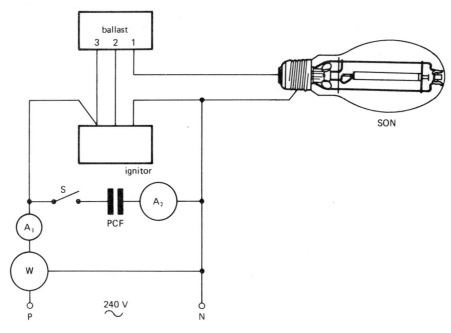

Fig. 6.24 SON discharge lamp circuit.

CHAPTER SEVEN

Instrument connections

After reading this chapter you will be able to:

1 Describe the operation of the following instruments:
   a) moving-coil instrument
   b) moving-iron instrument
   c) dynamometer wattmeter
   d) power-factor meter

2 Draw diagrams of the instruments listed.

3 Draw circuit diagrams of wattmeter connections for single-phase and three-phase supplies.

4 State the use of the instruments listed as well as current and voltage instrument transformers.

5 Perform calculations to determine values of shunt and multiplier resistors.

6 Perform calculations to determine power and power factor in single-phase and three-phase supply systems.

Indicating instruments

The electrical industry is flooded with many different types of measuring instrument and some of these have very complex operation. Today's choice is often made between analogue and digital types with the latter increasing on the market at a tremendous rate. This chapter, reflecting course schemes, is limited to mentioning briefly some of the traditional and modern instruments used for measuring resistance, current, voltage, power and power factor.

Instruments fitted with a pointer are called *analogue* instruments since their measurement operation relies on some form of physical quantity. An indicating instrument has a moving system which carries a pointer and it will be characterized by having three torques, namely, (i) a *deflecting torque* which moves the pointer over the instrument's scale as a result of the quantity that is being measured, (ii) a *controlling torque* which controls the amount of deflection and allows the needle to stop so that the scale can be read and it also allows the pointer to return to its zero position when the instrument is disconnected, and (iii) a *damping torque* to prevent

the needle from oscillating, allowing it to indicate its final steady value as quickly as possible.

In practice, nearly all direct-acting indicating instruments are operated by current. The deflecting torque is proportional to the current and to the magnetic flux density. The two most common methods used for the controlling torque are *hairspring* and *gravity*. The former provides the more accurate control since it is proportional to the angular deflection of the pointer. Two hairsprings are used and they are wound in opposite directions so that when the moving system is deflected, one spring winds up while the other unwinds. The latter control uses weights but does not provide the same degree of accuracy – the total deflection is limited to approximately 80%.

Damping control is achieved by *air dashpot* or *eddy current*. An air dashpot comprises a vane attached to the moving system which is made to move inside a sector-shape container. Air is compressed on one side of the container and this has the effect of damping any oscillations that occur. The other method achieves damping by allowing induced currents in part of the moving system and it results in a force being exerted which obeys Lenz's Law – i.e. the currents exert a force opposing the motion producing them. Figure 7.1 shows some of the methods used in the control and damping of indicating instruments.

Moving-coil instrument

This is a well-known, direct-acting indicating instrument commonly used for measuring current and voltage as well as resistance (see Figure 7.2). Such instruments are called galvanometers, ammeters, ohmmeters and voltmeters. The basic instrument relies on the d.c. 'motor principle' for its operation and consists of a rectangular coil of insulated copper wire wound on a light aluminium frame. It is pivoted and assembled on a fixed cylindrical iron core so that the frame is free to turn through almost 90° between the poles of a permanent magnet. The current being measured is lead into and out of the coil by spiral

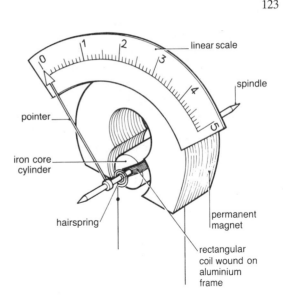

Fig. 7.2 Moving-coil instrument.

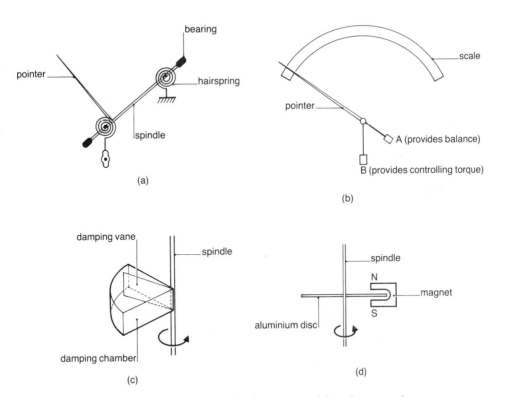

Fig. 7.1 Methods of control and damping in indicating instruments (a) spring control
(b) gravity control
(c) air damping
(d) eddy current damping.

hairsprings and the deflecting torque is produced by the interaction between the current-carrying coil magnetic field and the permanent magnet's magnetic field. Normally eddy current damping is provided and the scale is made *linear*. The instrument must be used *only* on direct-current supplies. Modern instruments are designed with a pivotless movement referred to as *taut band suspension* or alternatively they may be designed with a robust *centrepole* movement. This advance in design is mainly to protect the instrument against vibration and mechanical shock. The chief advantages with the moving-coil instrument are its high sensitivity, uniform scale and its magnetic shielding which protects it from any stray magnetic fields.

The moving-coil instrument is not capable of carrying any large current since milliampere sensitivity is quite common. In order, therefore, to extend the current or voltage range of this instrument, resistors called *shunts* and *multipliers* are used. The shunt resistor is a low ohmic value resistor which is connected in parallel with the instrument so that it shares a large proportion of the circuit current and allows the instrument to have its scale modified to the new circuit conditions. Ammeters use shunt resistors. The multiplier resistor is a 'dropper' resistor and in contrast is a high ohmic value resistor which is connected in series with the instrument so that it allows only a very small current to flow. It would therefore be used when the instrument scale was to be modified to read a high voltage such as with a voltmeter. Figure 7.3 shows shunt and multiplier

connections. The following example illustrates the method of determining the values of these two resistors.

Example

A moving-coil instrument gives full-scale deflection with 15 mA and has a moving assembly with a resistance of 5 Ω. Calculate the value of resistance to be connected (i) in parallel to enable the instrument to read a current of 3 A, (ii) in series with the instrument to enable it to be used as a voltmeter to read a voltage of 250 V.

Solution

For full-scale deflection, the instrument requires a voltage of

$$V = I \times R$$
$$= 0.015 \times 5 = 0.075 \text{ V (75 mV)}$$

From Figure 7.3, the current through the shunt resistor will be

$$I_s = 3 - 0.015$$
$$= 2.985 \text{ A}$$

Since the p.d. across the instrument will be the same as that across the shunt resistor, then the shunt's value is:

$$R_s = V/I_s$$
$$= 0.075/2.985$$
$$= \underline{0.025 \ \Omega}$$

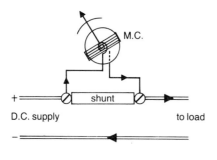

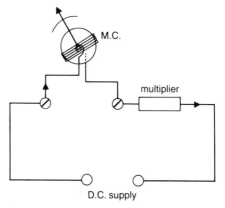

Fig. 7.3   Shunt and multiplier connections.

For use as a voltmeter connected to a 250 V supply, the instrument requires only 75 mV for f.s.d. This means that the dropper resistor has to have a p.d. of:

$$V_m = V_s - V_i$$
$$= 250 - 0.075$$
$$= 249.925 \text{ V}$$

Since only 15 mA is required in the series circuit for f.s.d. then the value of the multiplier is:

$$R_m = V/I$$
$$= 249.925/0.015$$
$$= 16661.67$$
$$= \underline{16.7 \text{ k } \Omega}$$

Note: I_s is shunt current

R_s is shunt resistor

R_m is multiplier resistor

V_m is the p.d. across multiplier

V_i is the p.d. across instrument

V_s is the supply voltage

The moving-coil instrument can be modified for use on a.c. supplies by designing it with a full-wave bridge rectifier. The a.c. is rectified into d.c. and the current flowing through the coil will measure the mean value. The scale is modified to read r.m.s. values.

The moving-coil instrument principle is also used in the operation of an ohmmeter which was discussed in the book *Theory and Regulations* by the same author.

Moving-iron instrument

There are two basic types of moving-iron instrument, namely, *attraction* type and *repulsion* type. The former type comprises a soft iron disc which is not centrally pivoted to its pointer. It is attracted into the instrument's coil by the magnetic field produced when the coil carries current. The latter type is more common and comprises fixed and moving irons situated inside the instrument's coil as shown in Figure 7.4. When current flows in the coil, both irons are magnetized simultaneously and at each end they acquire the same magnetic polarity. As a result both irons repel each other and the moving iron will indicate the magnitude of the current taken.

The value of the deflecting torque with both types of instrument is proportional to the square of the current (providing saturation is not reached) and it means that the scale will be cramped at both the low reading end and high reading end. This can be overcome in the repulsion instrument by using a scroll-type of fixed iron (see Figure 7.5). The chief advantages with these instruments are that they can be used on both a.c. and d.c. supplies and that they are robust as well as having a more favourable cost comparison with the moving-coil instrument. On a.c. supplies, the instrument will read r.m.s. values of current or voltage.

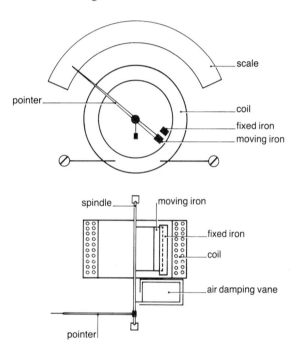

Fig. 7.4 Moving-iron instrument.

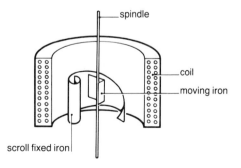

Fig. 7.5 Scroll-type fixed iron.

The dynamometer wattmeter

This instrument is used for measuring *power* and its operation depends upon the electromagnetic force exerted between fixed and moving coils when they are carrying current. Figure 7.6 shows the construction of this instrument while Figure 7.7 shows the magnetic field interaction between the coils. Briefly, the main field is produced by two fixed *current coils* and the moving *voltage coil* is pivoted between them. A high-value, non-inductive *swamping* resistor is connected in series with the voltage coil to ensure that the inductive reactance of the winding is of negligible value and this also allows the current through the voltage coil to be proportional and almost in-phase with the supply voltage. Thus, it will be seen that the deflection is proportional to the product of the two currents passing through the fixed and moving coils ($P = V \times I$).

The instrument can be used on both a.c. and d.c. supplies since any change of direction of current in the circuit affects both coils and the direction remains unaltered. The wattmeter measures the true power (P) in the circuit and this means that deflection is proportional to the voltage, current and power factor. Its scale is not quite linear and the moving system takes up a position depending on the mean torque. The instrument therefore reads the mean power. Small errors can be introduced in the instrument since power losses occur in the current coil and voltage coil. The instrument should be connected as in Figure 7.6(b).

Note: Modern types of wattmeter are electronically operated and some of these are portable, hand-held digital types capable of measuring not only power but also voltage, current and power factor.

Power-factor meter

This type of indicating instrument can be of the dynamometer type or moving-iron type. Both instruments are shown in Figure 7.8. The former type is used on single-phase supplies and has two fixed current coils in series with the load and a moving system comprising two coils at 90° to each other. Briefly, if a resistive component were being measured, the coil marked X would line up with the

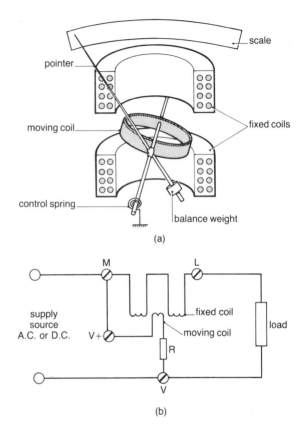

(a)

(b)

Fig. 7.6 Dynamometer wattmeter (a) construction
(b) circuit diagram.

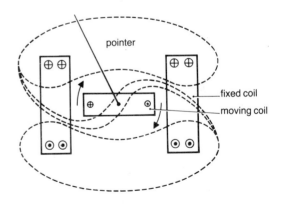

Fig. 7.7 Magnetic field interaction in a dynamometer wattmeter.

magnetic field of the fixed coils and the pointer would read *unity power factor*. When an inductive component is connected in circuit, coil Y lines up with the field from the fixed coils, swinging anticlockwise. This will represent a *lagging power factor* on the scale. When a capacitive component is connected in circuit, coil Y swings in a clockwise direction and these will represent a *leading power factor* on the scale. Deflection of the moving system is a result of the two torques created by magnetic field interaction.

In the moving-iron type, the moving element consists of an iron rotor carrying a pointer. When it is magnetized by a stationary voltage coil, the rotor moves within a rotating magnetic field created by the stationary current coils. The torque exerted by the rotating magnetic field causes the rotor to take up a position where the rotor field is in phase with the rotating field. In this position the pointer indicates on the scale the phase angle between the voltage and current of the circuit. A *Phase rotation indicator* operates on this principle and uses three coils in order to establish an automatic red, yellow and blue phase sequence from the supply. This instrument is used to determine the correct supply sequence (RYB) and its rotor will denote this by rotating in either a clockwise or anticlockwise direction. Wattmeters and power-factor meters are also designed to operate on three-phase supplies.

Figure 7.9 is an illustration of a modern electronic power-factor meter. With this instrument there is no need to interrupt the supply since it is a clamp-on type meter. It needs to have a source of supply which is obtained from the heavily insulated voltage probes. The instrument is suitable for single-phase and three-phase supplies and can also measure currents between 3 A and 500 A and voltages between 100 V and 600 V.

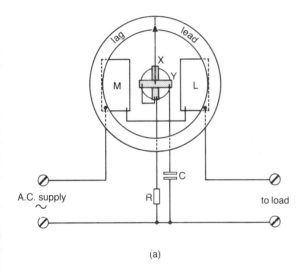

(a)

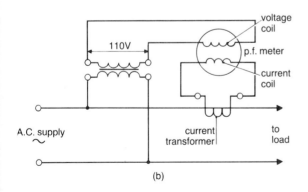

(b)

Fig. 7.8 Single-phase power factor meters
      (a) dynamometer
      (b) moving-iron type.

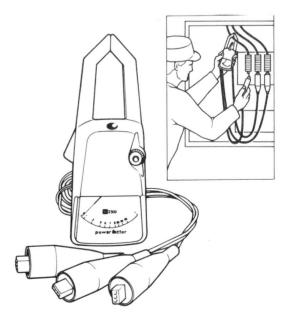

Fig. 7.9 Portable power factor meter.

Measurement of power and power factor

In d.c. supply systems, measurement of power is quite easily found by using a single wattmeter of the type described above or alternatively found by connecting a voltmeter and ammeter in the circuit and multiplying their readings together. Students must remember that ammeters need to be connected in series with the load and they must be capable of reading the full circuit design current. Voltmeters are normally connected across pressure points in a circuit in order to measure potential difference or e.m.f. Where portable instruments are used, it should be noted that many are capable of measuring other quantities and care should be taken to select the correct range and quantity wishing to be measured.

The above comments apply equally well to a.c. systems but in these systems the ammeter and voltmeter will measure only the *apparent power* (VA). The connection of a wattmeter will measure the *true power* (W). Figure 7.10(a) shows a diagram of the three instruments and their readings will enable the circuit's power factor to be found, i.e.

Power factor = wattmeter reading/(voltmeter reading × ammeter reading)

Where the current is very large and/or supply voltage very high, then instrument transformers will need to be used as shown in Figure 7.10(b). Some safety precautions were described in an earlier chapter but the diagrams show the importance of earthing one side of the secondary circuit of the instrument transformers and this includes transformer cores and any instrument casing which is an exposed conductive part. It is also important to remember that wattmeters and other instruments may have *scale multiplying factors* and any readings taken must consider this modification.

In three-phase systems, the method of measuring power using wattmeters can involve several forms of connection and often depends on whether or not the load is balanced or unbalanced. Figure 7.11 shows how one wattmeter is used for measuring the power in a 3-phase, 4-wire system when the load is balanced. The total power is three times the wattmeter reading. Where the system is 3-phase, 4-wire and unbalanced then three wattmeters in each line are used, their sum being the total power. This

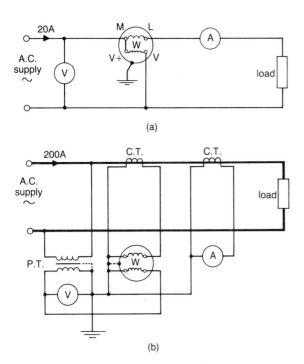

Fig. 7.10 Instrument connections (a) low current (b) high current.

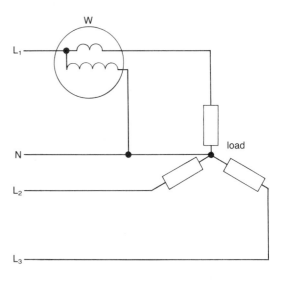

Fig. 7.11 One wattmeter method.

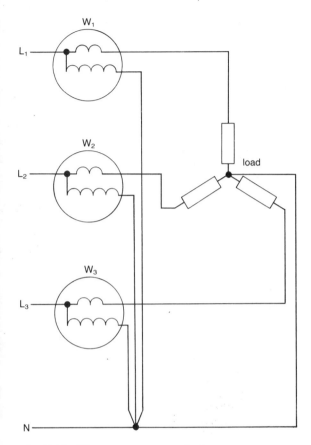

Fig. 7.12 Three wattmeter method.

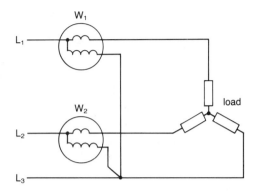

Fig. 7.13 Two wattmeter method.

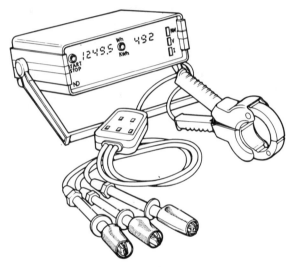

Fig. 7.14 Portable power meter.

method is shown in Figure 7.12. An arrangement which is used for measuring power in a 3-phase, 3-wire system which can be either balanced or unbalanced is shown in Figure 7.13 and it will be seen that use is made of two wattmeters. The total power is the sum of the two wattmeter readings. If the load is balanced, power factor can be found from the formula $\tan \phi = \sqrt{3}(W_2 - W_1)/(W_2 + W_1)$. By finding the phase angle ϕ, the power factor ($\cos \phi$) can be obtained.

Figure 7.14 is an illustration of a *portable power meter* designed for testing three-phase equipment ranging from 5 A to 1500 A. It operates on both a.c. and d.c. supplies and measures power, voltage and current; it also records energy on its digital display.

Exercise 7

1 Draw a circuit diagram showing how a single-phase wattmeter, voltmeter and ammeter are connected to a resistive load. Assume the use of a current transformer in view of the load current being high.

2 A moving-coil instrument has a resistance of 5 Ω and gives f.s.d. with a current of 15 mA. Calculate the value of the multiplier required to extend the voltage range to: (i) 1 V, (ii) 5 V, (iii) 15 V.

3 *a)* Draw the circuit diagram showing how a potential transformer and current transformer are connected in a single-phase supply system so that a wattmeter can be used to measure power.

   b) Briefly state some advantages of using instrument transformers in a.c. systems.

4 Draw a circuit diagram showing how an ammeter and selector switch are connected via current transformers in a three-phase, three-wire system in order for the ammeter to measure current in any line.

5 During a test on a single-phase motor the following readings were taken: voltmeter 240 V, ammeter 97.5 A and wattmeter 17.55 kW. What is the power factor of the motor?

6 A wattmeter is connected in one of the phases of a three-phase, four-wire a.c. system. The supply feeds a balanced load and the wattmeter reads 30 kW. Determine the current in each supply conductor if the line voltage is 400 V. What is the current in the neutral conductor?

7 Show how a voltmeter and selector switch are connected together to show the phase voltage in a three-phase, four-wire system.

8 *a)* Describe with the aid of a sketch a method of 'damping' used to prevent an instrument's pointer from vibrating.

   b) Explain the use of a 'shunt' resistor.

9 What are the necessary precautions to be taken when disconnecting an instrument from a current transformer? Show a circuit diagram of a method of achieving disconnection safely.

10 Two single-phase wattmeters are used to measure the power in a three-phase balanced system. Their readings are: W1 = 50 kW and W2 = 64 kW. Determine (i) the total power (ii) the power factor of the system.

CHAPTER EIGHT

Revision exercise

At the end of this revision exercise you will be able to:

1 Apply laws and formulae to a number of specific areas associated with City & Guilds Part I and II electrical schemes.

2 Perform numerous calculations relating to basic circuit theory at Part I Certificate level.

3 Perform numerous calculations relating to basic circuit theory at Part II Certificate level.

Part I Examination Syllabus

1 Two resistors of 4 Ω and 12 Ω are connected in parallel and they are both joined to a further resistor of 10 Ω connected in series. Draw the circuit diagram. If a d.c. supply of 78 V is connected to the circuit, determine:
   a) the total circuit resistance
   b) the total current
   c) the potential difference across the parallel resistors and the potential difference across the series resistor
   d) the branch currents in the parallel circuit
   e) the quantity of electricity used after five hours
   f) the total power consumed
   g) the power consumed by the parallel resistors
   h) the power consumed by the series resistor
   i) the total energy used after a period of five hours
   j) the cost after 65 hours of use at 5.16 p unit

Answers: (a) 13 Ω, (b) 6 A, (c) 18 V and 60 V, (d) 4.5 A and 1.5 A, (e) 108 kC, (f) 468 W, (g) 108 W, (h) 360 W, (i) 6.48 MJ, (j) £3.35.

2 A 3 µF and 6 µF capacitor are connected in series and joined in series to two 5 µF capacitors connected in parallel. Draw the circuit diagram. If the supply voltage is 30 V d.c., determine:
   a) the total capacitance
   b) the total charge
   c) the potential difference across each series capacitor
   d) the potential difference across the parallel capacitors

   e) the total energy stored

Answers: (a) 1.67 µF, (b) 50 µC, (c) 16.67 V and 8.33 V, (d) 5 V, (e) 751.5 µJ

3 A 2.5 mm² pvc/pvc/cpc copper cable is used to feed a 3 kW/240 V immersion heater circuit. The length of run is 25 m and the resistivity of copper is 0.0172 µΩ m. Determine:
   a) the resistance of the cable
   b) the voltage drop in the circuit
   c) the maximum voltage drop allowed based on 2.5% of the supply voltage
   d) the terminal voltage at the heater
   e) the actual power consumed by the heater.

Answers: (a) 0.344 Ω, (b) 4.3 V, (c) 6 V, (d) 235.7 V, (e) 2.89 kW

4 A single-phase, double-wound transformer has a step-up ratio of 1:1.73 and has a primary winding consisting of 400 turns. If its primary current is 8 A and its secondary voltage is 240 V, determine:
   a) the primary voltage
   b) the secondary current
   c) the secondary turns
   d) the volts per turn
   e) the load kVA

Answers: (a) 138.7 V, (b) 4.62 A, (c) 692 turns, (d) 0.347 V, (e) 1.11 kVA

5 An electric kettle is rated at 3 kW/240 V and contains 1 litre of water. If the change of water temperature is 80°C from cold to boiling point and the time taken to boil is two minutes, determine:
   a) the heat energy required
   b) the electrical energy

c) the percentage efficiency
Note: the specific heat capacity of water is 4.2
kJ/kg °C and 1 litre of water is equivalent to 1 kg.
Answers: (*a*) 0.336 MJ, (*b*) 0.1 kWh, (*c*) 93%

6   With reference to the balanced system shown in
Figure 8.1, determine:
   a)   The phase current in the delta connection
   b)   the phase current in the star connection
   c)   the phase voltage in the delta connection
   d)   the phase voltage in the star connection
   e)   the ratio between the line and phase
        voltages in star.

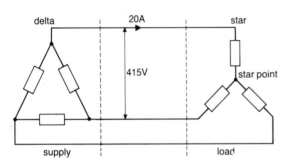

Fig. 8.1   Delta-star connections.

Answers: (*a*) 11.55 A, (*b*) 20 A, (*c*) 415 V,
             (*d*) 240 V, (*e*) $V_L = \sqrt{3}V_p$

7   *a*)   A 2 V secondary cell is accidently short-
          circuited. If the p.d. of the cell is 1.2 V and
          the short circuit current is 200 A, determine
          the internal resistance of the cell.
    b)   If the secondary cell is 90% efficient and
          discharges 3 A for 20 hours, how long will it
          take to recharge using a current of 4 A?
Answers: (*a*) 0.004 Ω, (*b*) 16.67 hours

8   An earth fault loop impedance test on a 240 V
    installation protected by a 30 A, BS 3036 fuse
    gives a reading of 6 Ω. On further investigation a
    metal conduit joint is found to have a resistance
    of 4 Ω. If a direct fault occurs between phase
    conductor and the conduit, determine:
    a)   the leakage current
    b)   the power developed at the joint
    c)   the voltage across the joint

Answers: (*a*) 40 A, (*b*) 6.4 kW, (*c*) 160 V

9   With reference to the impedance triangle shown
    in Figure 8.2:
    a)   Determine the impedance (Z).
    b)   If R remains at 20 Ω and X_L were reduced
         to 20 Ω, what is the new value of Z?
    c)   What would be the power factor of the
         circuit if $X_L = 0$?
    d)   What component would reduce X to zero if
         it represented inductive reactance?

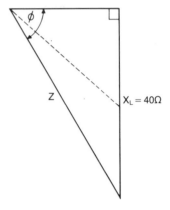

Fig. 8.2   Impedance triangle.

Answers: (*a*) 44.72 Ω, (*b*) 28.28 Ω, (*c*) unity, (*d*)
             capacitor

10   Figure 8.3 shows a sinusoidal alternating voltage
     quantity over one complete cycle:
     a)   Express the periodic time in terms of
          frequency.
     b)   How many times will the a.c. cycle be
          extinguished in 1 cycle?
     c)   What term is used to express the highest
          point reached during any one half-cycle?
     d)   If the sinewave quantity had an r.m.s. value
          of 240 V, what would be its maximum
          value?
     e)   What is the mean value of the sinewave
          quantity over (i) one half-cycle, (ii) one
          complete cycle?
Answers: (*a*) $T = 1/f$, (*b*) twice, (*c*) maximum or
             peak value, (*d*) 339.46 V, (*e*) (i) 0.637, (ii)
             zero

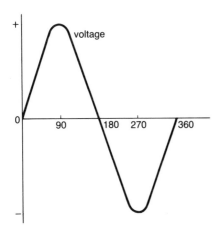

Fig. 8.3 Alternating voltage quantity.

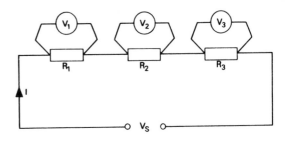

Fig. 8.4

Multiple-choice questions—Part I Certificate

1 The quantity of electricity (Q) is found by the expression
   a) $Q = V \times I$
   c) $Q = I \times t$
   b) $Q = P \times t$
   d) $Q = I \times R$

2 The unit of reactance is called the
   a) volt
   c) watt
   b) ampere
   d) ohm

3 If the charge conveyed in an electric circuit is 300 C, the time taken to pass a current of 3 A is
   a) 100 s
   c) 303 s
   b) 297 s
   d) 900 s

4 In Figure 8.4, the potential difference across R_1 is found by
   a) $V_1 = V_s - V_2 - V_3$
   b) $V_1 = V_s + V_2 + V_3$
   c) $V_1 = V_s - V_2 + V_3$
   d) $V_1 = V_s + V_2 - V_3$

5 What is the power dissipated by a 10 Ω resistor supplied at 200 V?
   a) 2 kW
   c) 10 kW
   b) 4 kW
   d) 20 kW

6 The current taken by a 60 W, 240 V tungsten lamp is
   a) 100 mA
   c) 250 mA
   b) 220 mA
   d) 400 mA

7 A double-wound transformer delivers 500 V to a load taking 500 kVA. What current is taken by the load?
   a) 100 A
   c) 1000 A
   b) 500 A
   d) 5000 A

8 A conductor 100 metres long has an insulation resistance of 50 megohms. What will be its insulation resistance for a length of 500 m?
   a) 250 megohm
   c) 25 megohm
   b) 120 megohm
   d) 10 megohm

9 Which of the following materials has a negative temperature coefficient of resistance?
   a) carbon
   c) silver
   b) brass
   d) copper

10 In Figure 8.5, the line-to-neutral voltage (V_n) is approximately
   a) 110 V
   c) 64 V
   b) 84 V
   d) 55 V

11 In Figure 8.6 the load phase current is
   a) 90 A
   c) 30 A
   b) 45 A
   d) 20 A

12 In Figure 8.6 the load line voltage is
   a) 415 V
   c) 240 V
   b) 320 V
   d) 110 V

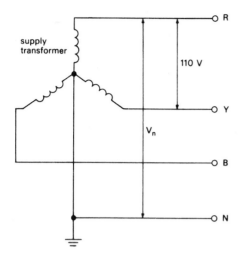

Fig. 8.5

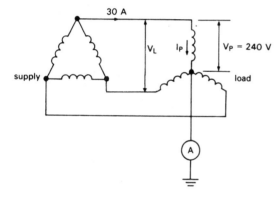

Fig. 8.6

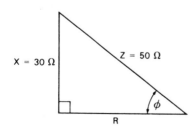

Fig. 8.7

13   In Figure 8.6, if all line conductors carried 30 A,
     the neutral ammeter would read
     a)   90 A   c)   10 A
     b)   30 A   d)   0 A

14   In Figure 8.7 the resistance and power factor
     respectively are found to be
     a)   45 Ω and 0.75   c)   35 Ω and 0.85
     b)   40 Ω and 0.80   d)   31 Ω and 0.90

15   The power factor of an a.c. circuit is given by
     the ratio

     a) $\dfrac{\text{power output}}{\text{power input}}$

     b) $\dfrac{\text{wattless power}}{\text{reactive power}}$

     c) $\dfrac{\text{true power}}{\text{apparent power}}$

     d) $\dfrac{\text{average power}}{\text{maximum power}}$

16   Balancing single-phase loads on three-phase,
     four-wire systems is to ensure that
     a)   line voltages are all equal
     b)   minimal neutral current flows
     c)   star point is maintained at all times
     d)   circuit fuses operate efficiently

17   In Figure 8.8, the system would be ideally
     balanced if the yellow phase load X carried
     a)   30 A   c)   64 A
     b)   50 A   d)   82 A

18   Which of the following is not recognized as a
     standard a.c. supply voltage in the UK?
     a)   11 000 V   c)   415 V
     b)   650 V   d)   240 V

19   The number of units of electricity read by the
     dials shown in Figure 8.9 are
     a)   95 794   c)   94 795
     b)   95 695   d)   94 694

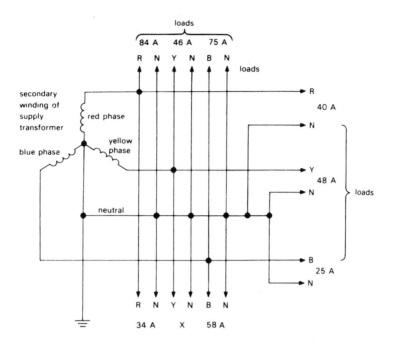

Fig. 8.8

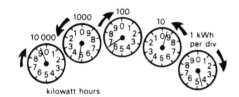

kilowatt hours

Fig. 8.9

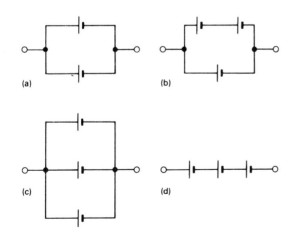

Fig. 8.10

20 The use of a capital M preceding a unit or unit symbol denotes
   a) metre   *c)* mega
   b) micro   *d)* milli

21 Which combination of 2 V cells in Figure 8.10 provides the highest voltage output?
   (*a*) (*b*) (*c*) (*d*)

22 The positive plate of a lead–acid cell on discharging changes to
   a) lead peroxide   *c)* spongy lead
   b) lead sulphate   *d)* water

23 What percentage of 110 V is 2.75 V?
   a) 10.0%   *c)* 2.5%
   b) 6.0%   *d)* 1.5%

24 In Figure 8.11 the instruments connected to the circuit may be used to determine
   a) energy   *c)* power factor
   b) efficiency   *d)* frequency

25 In a three-phase, star-connected circuit, the ratio $\dfrac{\text{line volts}}{\sqrt{3}}$ gives

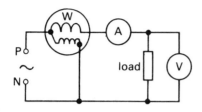

Fig. 8.11

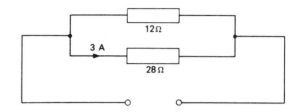

Fig. 8.13

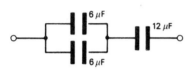

Fig. 8.12

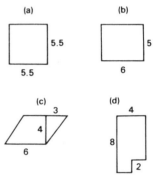

Fig. 8.14

a) the voltage to neutral
b) the voltage across two lines
c) the permissible volt drop
d) neutral/earth voltage

26 The overall value of capacitance in Figure 8.12 is
    a) 15 μF   c) 6 μF
    b) 12 μF   d) 3 μF

27 If the capacitors shown in Figure 8.12 were
    connected to supply of 500 V the charge
    conveyed in the circuit would be
    a) 3 MC   c) 3 mC
    b) 3 kC   d) 3 μC

28 In Figure 8.13, the power dissipated in the 12
    ohm resistor is
    a) 604 W   c) 433 W
    b) 588 W   d) 309 W

29 Given $S = \dfrac{\sqrt{I^2 t}}{k}$, what is the value of S
    when $I = 400$ A, $t = 0.035$ s, and $k = 115$?
    a) 1.00 mm²   c) 0.75 mm²
    b) 0.82 mm²   d) 0.65 mm²

30 To express kWh in basic SI units, 1 kWh is equal
    to
    a) 3.6 MJ   c) 3.6 mJ
    b) 3.6 kJ   d) 3.6 μJ

31 Which of the following distribution voltages is
    standard in the UK?
    a) 450 V   c) 415 V
    b) 440 V   d) 440 V

32 In Figure 8.14 which shape has the greatest area?
    (a)   (b)   (c)   (d)

33 In a factory supplied at 415 V, three-phase four-
    wire, the phase voltage would be
    a) 415 V   c) 100 V
    b) 240 V   d) 64 V

34 How long will it take a 2.5 kW water heater to
    use 5 MJ of energy?
    a) 5000 s   c) 1250 s
    b) 2000 s   d) 500 s

35 The previous and present energy meter readings in a domestic dwelling are 44157 and 45409 respectively. If the standing charge per quarter is £5.07, the consumer's quarterly bill at 4.18p per unit, is
a) £67.46   *c)* £43.25
b) £57.40   *d)* £31.87

36 An alternating e.m.f. has a frequency of 50 Hz. What is the time over one cycle?
a) 5.00 s   *c)* 0.25 s
b) 2.50 s   *d)* 0.02 s

37 The ampere-hour efficiency of a secondary cell is 80%. If the Ah on discharge is 50, the Ah charge is
a) 85.0   *c)* 62.5
b) 74.5   *d)* 57.0

38 The specific heat capacity of water is
a) 5800 J/kg °C   *c)* 3600 J/kg °C
b) 4200 J/kg °C   *d)* 2000 J/kg °C

39 What is the maximum value reached by an a.c. supply of 415 V?
a) 931 V   *c)* 654 V
b) 829 V   *d)* 587 V

40 The permitted voltage drop allowed on consumers' wiring supplied at 110 V is
a) 6.00 V   *c)* 2.75 V
b) 5.25 V   *d)* 1.33 V

41 In Figure 8.15, to obtain a resistance of 75 Ω the switch positions would be
a) S1 off, S2 on, S3 on
b) S1 on, S2 on, S3 off
c) S1 off, S2 on, S3 off
d) S1 on, S2 off, S3 on

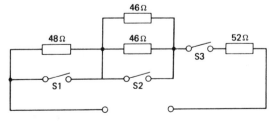

Fig. 8.15

42 The current which flows through a 1 Ω resistor connected to a primary cell of 1.4 V having an internal resistance of 0.4 Ω is
a) 2.0 A   *c)* 1.0 A
b) 1.4 A   *d)* 0.4 A

43 The energy possessed by a body due to its motion is called
a) electrical energy
b) thermal energy
c) kinetic energy
d) potential energy

44 The rate at which work is done is called
a) energy   *c)* force
b) power   *d)* acceleration

45 When two closely parallel conductors pass current in opposite directions, the effect will be to
a) repel each other
b) attract each other
c) rotate about each other
d) short-circuit each other

46 The volume of a copper rod 1 m long and 2 mm in diameter is
a) 3142 mm³   *c)* 31.42 mm³
b) 314.2 mm³   *d)* 3.142 mm³

47 A transformer has a primary-to-secondary turns ratio 25:1. If the primary voltage is 250 V, then the secondary voltage is
a) 6.25 kV   *c)* 25 V
b) 250 V   *d)* 10 V

48 Ignoring temperature change of the element, when a 3 kW, 240 V immersion heater is supplied at 120 V its power is reduced to
a) 1500 W   *c)* 750 W
b) 1000 W   *d)* 250 W

49 The coulomb is the unit of
a) magnetic flux
b) magnetomotive force
c) electric charge
d) capacitance

50 On a 50 Hz supply, a fluorescent tube
    extinguishes itself
    a) 200 times every second
    b) 100 times every second
    c) 50 times every second
    d) twice every second

Part II Examination Syllabus

1 *a*) An inductive coil has resistance, reactance
        and impedance when connected to an a.c.
        supply. State which of these effects will
        increase with increase in frequency.
   b) For the circuit shown in Figure 8.16 draw
        the phasor diagram and calculate the
        (i)   impedance
        (ii)   current
        (iii) power factor
        (iv)   phase angle
        (v)   power

                                        CGLI/II/84

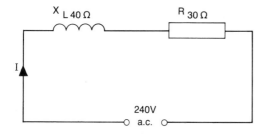

Fig. 8.16

Answers: (*a*) Each component represents an
           opposition to current but reactance
           increases with frequency $(X_L = 2\pi f L)$.
           Since impedance $Z = \sqrt{(R^2 + X^2)}$, this
           also increases. (*b*) Draw a phasor diagram
           with current quantity as the reference
           (i) $Z = 50\ \Omega$, (ii) $I = 4.8$ A, (iii) 0.6
           lagging, (iv) 53°, (v) 691.2 W

2 A moving-coil milliammeter has a resistance of
   5 Ω with full-scale deflection at 15 mA. The scale
   reads 0–15 mA with 15 divisions.
   a) With the aid of circuit diagrams explain how
        (i)   the current range of the instrument
             may be extended
        (ii) the milliammeter can be adapted for
             use as a voltmeter.

b) (i)   Calculate the value of the resistor
           required to enable the milliammeter to
           read 0–3 A.
      (ii) By what factor must the scale reading
           be multiplied to give correct readings
           or current?
c) (i)   Calculate the value of the resistor
           required to enable the milliammeter to
           read 0–150 V.
      (ii) How many volts will one division of
           the scale now represent?
                                        CGLI/II/80
Answers: A full solution to this question is given on
           page 72, *WEEI*. (*a*) (i) use of shunt
           resistor (ii) use of series resistor,
           (*b*) (i) 0.025 Ω, (ii) 200,
           (*c*) (i) 9.995 kΩ, (ii) 10

3 *a*) Explain what is meant by the 'power factor'
        of an a.c. circuit.
   b) State *two* disadvantages of a low power
        factor.
   c) The rated output of a single-phase
        transformer is 6 kVA, 240 V.
        Calculate
        (i)   the full-load current
        (ii)   the full-load power at unity power
              factor
        (iii) the full-load power and reactive volt-
              amperes at 0.8 power factor lagging.
                                        CGLI/II/87
Answers: (*a*) A.C. current and voltage are not
           always in phase with each and a factor
           called 'power factor' (p.f.) is used as a
           multiplier to the power equations. For
           example, $P = VI \times$ p.f. Re-arranging the
           equation, p.f. $= P/(VI)$. This ratio is $\cos\phi$
           where ϕ is the phase angle and lies
           between 0 and 90°C. Thus, power factor
           $(\cos\phi)$ ranges between 1 (unity) and zero.
           (*b*) A large phase angle means a poor
           power factor since it draws more current
           from the supply. Two disadvantages are:
           larger circuit cables and larger switchgear,
           both leading to higher installation costs
           (*c*) (i) 25 A, (ii) 6 kVA, (iii) 4.8 kW,
           3.6 kVAr

4 *a*) With the aid of sketches show how a

voltmeter, an ammeter and a wattmeter may be connected, using suitable transformers, to an 11 kV single-phase circuit.

b) A moving-coil voltmeter has a resistance of 400 kΩ and a full-scale deflection of 20 V. Calculate the value of a resistor, connected in series, which would allow the instrument to measure up to 250 V d.c.

CGLI/II/87

Answers: (a) See Figure 8.17, (b) 4.6 MΩ

5 A 240 V, 50 Hz, 6-pole split-phase motor is to be used to drive a pump required to lift four litres of water per second through a height of 10 m. The pump and motor have efficiencies of 74% and 80% respectively with a motor power factor of 0.85 lagging. Take g as 9.81 m/s^2.
a) Calculate the
   (i) input power
   (ii) current
b) State *three* requirements of the *IEE Regulations* for the control of electrical motors above 0.375 kW.

CGLI/II/82

Answers: (a) (i) 663 W, (ii) 3.24 A, (b) See IEE Regulations, section 552

6 A lamp placed 4 m above a road has a luminous intensity of 1600 cd in the direction LP on the road surface as shown in Figure 8.18.
a) Calculate the illuminance at the point P on this figure.

b) What is meant by the terms:
   (i) luminous flux
   (ii) utilization factor
   (iii) maintenance factor
   (iv) efficacy

CGLI/II/82

Answers: (a) 51.2 lx (b) See 'Theory & Regs' book.

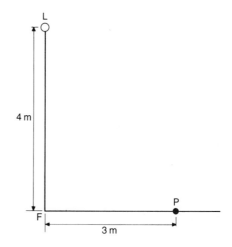

Fig. 8.18

7 a) Explain how a capacitor is used to provide the starting torque of a single-phase induction motor.
b) A 4-pole induction motor has a full-load speed of 23 rev/s when running from a 50 Hz supply. Calculate the percentage slip.
c) A 240 V single-phaser motor has a full-load output of 950 W at a power factor of 0.75 and an efficiency of 60% Calculate the line current at this load.

CGLI/II/86

Answers: (a) See Chapter 4 on electric motors, (b) 8%, (c) 8.8 A

8 a) Explain what is meant by the term impedance.
b) When connected to a 40 V d.c. supply, an iron-cored coil takes a current of 2.5 A and when connected to 40 V, 50 Hz supply the current taken is 2 A. Calculate the value of:
   (i) resistance
   (ii) impedance

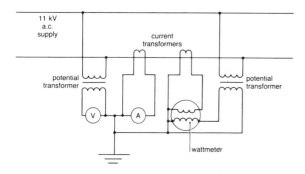

Fig. 8.17

(iii) reactance
(iv) power factor
c) Explain the effect upon the power factor of the coil if the iron core were removed.

CGLI/II/83

Answers: (a) Total opposition to current flow in a.c. circuits,
(b) (i) 16 Ω, (ii) 20 Ω, (iii) 12 Ω, (iv) 0.8 lagging,
(c) This reduces the strength of magnetic field in the coil and in turn reduces the amount of induced e.m.f. As a result, the current will not lag behind the voltage as much as it did previously and therefore the power factor will be improved.

9 a) Explain briefly how the back e.m.f. and the current change during the starting of a d.c. motor.
b) A 200 V shunt-motor has an armature resistance of 0.25 Ω and a field resistance of 200 Ω. The motor gives an output of 4 kW at an efficiency of 80%. For this load calculate:
(i) the motor input in kW
(ii) the load current
(iii) the motor field current
(iv) the armature current
(v) the back e.m.f.
c) What are the *IEE Regulations* for the rating of fuses protecting a circuit feeding a motor?

CGLI/II/1980

Answers: (a) See Chapter 4 on electric motors, (b)
(i) 5 kW, (ii) 25 A, (iii) 1 A, (v) 24 A,
(iv) 194 V, (c) See *IEE Regulations*, Section 552.

10 a) State *three* reasons why current transformers are often used in the metering systems of large installations.
b) Why are 'tong testers' (clamp-on-ammeters) usually preferred to other types of meter when it is desired to measure currents flowing in an installation?
c) A load has an expected demand of 200 kVA balanced across a 415 V 50 Hz, three-phase supply. The current is to be indicated by a 5 A meter arranged to measure the current in one line, using a current transformer.

(i) Determine the ratio of the current transformer
(ii) Draw a circuit diagram of the arrangement.

CGLI//II/1985

Answers: (a) See Chapter 5 on transformers and rectifying devices, (b) No need to disconnect circuits, (c) (i) 60:1 (ii) see Chapter 7 on instrument connections.

Multiple-choice questions—Part II Certificate

1 A conductor 15 mm in length, lies at right angles to a magnetic field of strength 15 T. What is the force exerted on it when carrying a current of 500 mA?
a) 112.5 MN   c) 112.5 mN
b) 112.5 kN   d) 112.5 μN

2 The per unit slip of an a.c. induction motor having a synchronous speed of 25 rev/s and rotor speed of 23.75 rev/s, is
a) 0.05   c) 0.02
b) 0.03   d) 0.00

3 In Figure 8.19, the current taken by the circuit is
a) 50 A   c) 10 A
b) 25 A   d)   5 A

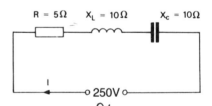

Fig. 8.19

4 The illuminance at point S in Figure 8.20 is
a) 25.60 lx   c) 6.40 lx
b) 17.30 lx   d) 5.12 lx

5 The power input of an electric pump which lifts 20 litres of water through a height of 5 metres in 12.26 seconds having an efficiency of 80% is, approximately
a) 500 W   c) 135 W
b) 250 W   d) 100 W

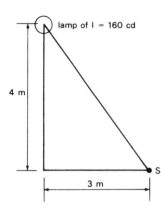

Fig. 8.20

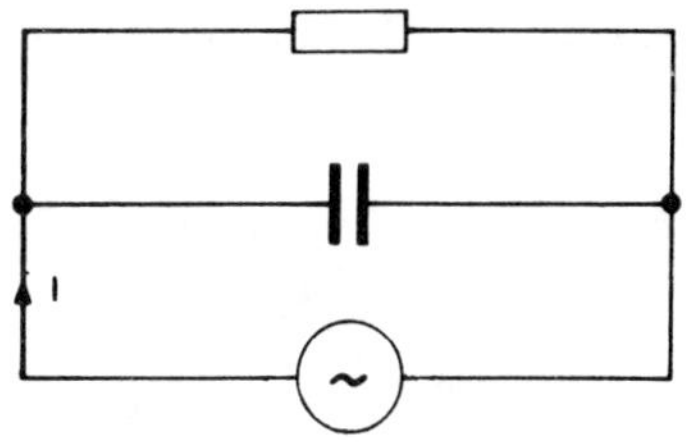

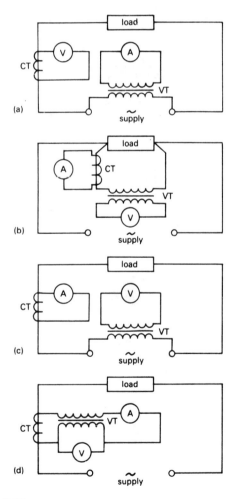

Fig. 8.21

6 A moving-coil instrument gives full-scale
   deflection with a current of 15 mA and has a
   resistance of 5 Ω. What is the value of a shunt to
   allow the instrument to read up to 5 A?
   a)   75 mΩ   *c)*   15 mΩ
   b)   42 mΩ   *d)*   5 mΩ

7 A single-phase, double-wound transformer
   having primary and secondary turns of 250 and
   50 respectively, takes a primary current of 20 A.
   What current is taken by the secondary circuit?
   a)   250 A   *c)*   50 A
   b)   100 A   *d)*   4 A

8 In Figure 8.21, if the resistor current was 6 A
   and the capacitor current was 8 A, what would
   be the current taken from the supply?
   a)   14 A   *c)*   8 A
   b)   10 A   *d)*   6 A

9 Which of the circuits in Figure 8.22 shows the
   correct method of connecting an ammeter and
   voltmeter through instrument transformers?
   (a) *(b)* *(c)* *(d)*

10 In an induction motor, per unit slip (*s*) is found
    by the expression
    a) $s = (n_r - n_s) \times n_s$
    b) $s = (n_s - n_r) \div n_s$
    c) $s = (n_r + n_s) - n_r$
    d) $s = (n_s + n_r) + n_r$

    where n_s is the synchronous speed
          n_r is the rotor speed

Fig. 8.22

11 In a d.c. motor, the back e.m.f. (E) is given by the expression

    a) $E = V - I_aR_a$ c) $E = V \div I_aR_a$
    b) $E = V + I_aR_a$ d) $E = V \times I_aR_a$

    where V is the supply voltage
            I_a is the armature current
            R_a is the armature resistance

12 The type of rotor shown in Figure 8.23 is called
    a) slip-ring   c) wound
    b) armature   d) cage

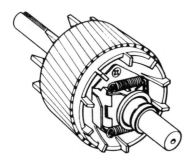

Fig. 8.23

13 The semiconductor device shown in Figure 8.24 is called a
    a) diac   c) transistor
    b) triac   d) thyristor

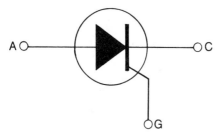

Fig. 8.24

14 What is the efficiency of a 250 V d.c. motor taking a current of 16.8 A if its rated ouptut is 3730 W?
    a) 88.8%   c) 66.7%
    b) 70.5%   d) 50.4%

15 Lamp efficacy is given by the ratio

    a) $\dfrac{\text{lumens}}{\text{watts}}$

    b) $\dfrac{\text{candelas}}{\text{metres}}$

    c) $\dfrac{\text{lux}}{4\pi r^2}$

    d) $\dfrac{\text{luminous intensity}}{\text{luminous flux}}$

16 For remote control of a direct-on-line contactor starter, start buttons are wired in
    a) series and stop buttons are wired in series
    b) parallel and stop buttons are wired in parallel
    c) series and stop buttons are wired in parallel
    d) parallel and stop buttons are wired in series

17 A device used to prevent a temperature rise in a semiconductor rectifier is called a
    a) thermocouple   c) rheostat
    b) heat sink   d) surge divertor

18 The measure of light falling on a surface is called
    a) luminance
    b) luminous flux
    c) luminous intensity
    d) illuminance

19 The ratio $\dfrac{\text{useful flux}}{\text{total emitted flux}}$ is called

    a) coefficient of utilization
    b) maintenance factor
    c) diversity factor
    d) room index

20 The unit of luminous intensity is the
    a) lux   c) candela
    b) lumen   d) nit

21 With reference to Figure 8.25, the illuminance at point A is approximately
    a) 133 lux   c) 44 lux
    b) 100 lux   d) 7 lux

22 With reference to Figure 8.25, the illuminance at point B is
    a) 44.4 lux   c) 9.6 lux
    b) 16.0 lux   d) 3.1 lux

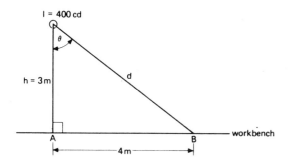

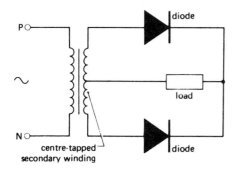

Fig. 8.25

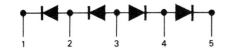

Fig. 8.26

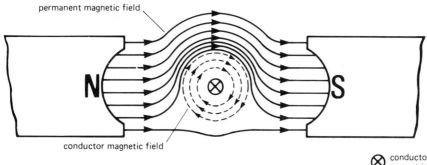

Fig. 8.27

23 Figure 8.26 shows the four elements of a bridge-connected rectifier. To which terminals will the a.c. supply be connected?
a) 1 and 3 . *c)* 3 and 5
b) 2 and 4 *d)* 4 and 1

24 In Figure 8.26 where would you connect the d.c. negative load terminal?
a) 1 *c)* 3
b) 2 *d)* 4

25 In Figure 8.26 to provide full-wave d.c., the terminals that require joining together are
a) 1 and 2 *c)* 1 and 4
b) 1 and 3 *d)* 1 and 5

26 What is the speed in rev/s of a four-pole induction motor running with 0.05 per unit slip and supply frequency 50 Hz?
a) 21.15 *c)* 24.25
b) 23.75 *d)* 25.00

27 Figure 8.27 represents a basic
a) thyristor inverter circuit
b) half-wave rectifier circuit
c) full-wave rectifier circuit
d) diode chopper circuit

28 In Figure 8.28 which direction will the conductor move?
a) left *c)* up
b) right *d)* down

![Figure 8.28 showing a conductor between magnetic poles N and S with field lines]

Fig. 8.28

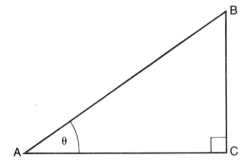

Fig. 8.29

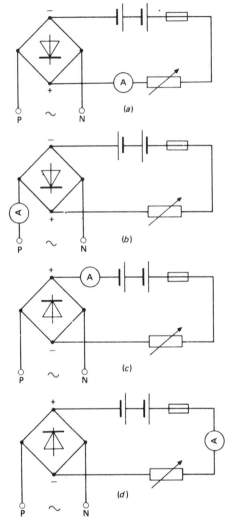

Fig. 8.30

29 The cosine of the angle in Figure 8.29 is the ratio
    of

a) $\dfrac{AB}{AC}$ c) $\dfrac{AC}{AB}$

b) $\dfrac{AB}{BC}$ d) $\dfrac{BC}{AC}$

30 The transformation ratio of a step-down
    transformer is 4.5:1. If the primary voltage is
    495 V its secondary voltage is
    a) 11 000 V   c) 240 V
    b)   415 V   d) 110 V

31 Which of the circuits in Figure 8.30 is correct for
    charging a secondary battery?
    (a)   (c)
    (b)   (d)

32 The ratio $\dfrac{\text{true power}}{\text{apparent power}}$

    is used for finding an a.c. circuit's
    a) reactive voltamperes
    b) load factor
    c) energy
    d) power factor

33 What is the wattage of a lamp if its efficacy is
    11 lm/W and its luminous flux is 275 lm?
    a) 10 W   c) 40 W
    b) 25 W   d) 60 W

34 In the radial distributor shown in Figure 8.31
    each section of the 2-core cable has a resistance
    of 0.01 Ω. The voltage at A is 240 V, the voltage
    at D is
    a) 240 V   c) 234 V
    b) 237 V   d) 231 V

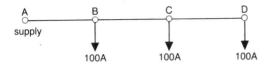

Fig. 8.31

35 In the circuit of Figure 8.32 the current (*I*) has a value of
  a) 30 A lagging   c) 10 A lagging
  b) 30 A leading   d) 10 A leading

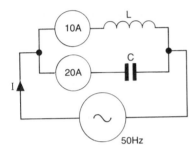

Fig. 8.32

36 The unit of illuminance is the
  a) lux   c) candela
  b) lumen   d) nit

37 The illuminance vertically below a lamp placed 3 m above floor level is 50 lux. If the lamp were raised by 2 m the illuminance at floor level would be
  a) 10 lux   c) 36 lux
  b) 18 lux   d) 42 lux

38 Which one of the circuits in Figure 8.33 could be used to measure the power in the load?
       (a) (b) (c) (d)

39 The normal precautions taken when an instrument is disconnected from a current transformer is to
  a) leave open a transformer's secondary winding
  b) short out the transformer's secondary winding

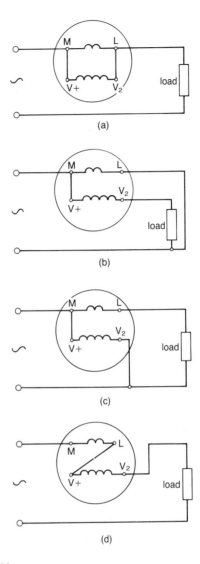

Fig. 8.33

  c) remove the transformer with the instrument
  d) replace the instrument with a winding lamp

40 In the shunt motor shown in Figure 8.34 the direction of rotation is clockwise. Which one of the circuit arrangements below will reverse this direction?
       (a) (b) (c) (d)

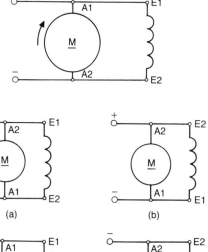

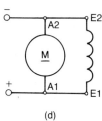

Fig. 8.34

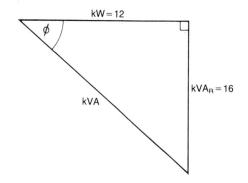

Fig. 8.35

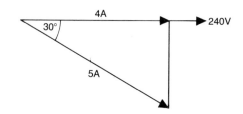

Fig. 8.36

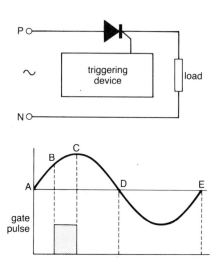

Fig. 8.37

41 The kVA represented in Figure 8.35 is

   a)   20   *c)*   28

   b)   25   *d)*   30

42 Figure 8.36 shows the phasor diagram of a single-phase motor. To improve the motor power factor to unity the capacitive reactance (X_c) needed would be

   a)   30 Ω   *c)*   70 Ω

   b)   50 Ω   *d)*   80 Ω

43 Copper or aluminium bars are used in the construction of a cage rotor induction motor to

   a)   provide a path for leakage current

   b)   conduct the heat away from the rotor

   c)   provide a path for induced current

   d)   reduce eddy current losses

44 The rotating field produced by the stator of a 3-phase induction motor travels at

   a)   a speed above synchronous speed

   b)   a speed below synchronous speed

   c)   the same speed as the rotor

   d)   synchronous speed

45 Figure 8.37 shows a thyristor connected in series with a load. On the a.c. graph, the thyristor will conduct only between the positions marked:

   a)   A to B   *c)*   B to D

   b)   A to C   *d)*   B to E

46 If a balanced load of 40 kW at a power factor 0.927 lagging is connected to 415 V, 50 Hz, 3-phase supply, the line current will be
 a) 60 A *c)* 212 A
 b) 104 A *d)* 240 A

47 The torque developed by 6 kW motor having a shaft speed of 25 rev/s is approximately
 a) 1 500 Nm *c)* 250 Nm
 b) 377 Nm *d)* 38 Nm

48 In Figure 8.38, the line current (I_L) is equal to
 a) $\sqrt{3}I_\mathrm{p}$ *c)* $\frac{1}{3}I_\mathrm{p}$
 b) $3I_\mathrm{p}$ *d)* $\sqrt{(3I_\mathrm{p})}$

49 The kVA input of a motor taking 4 kW and 3 kVAr
 a) 3 kVA *c)* 5 kVA
 b) 4 kVA *d)* 7 kVA

50 The motor terminal markings in Figure 8.39 identify it as a
 a) d.c. shunt motor
 b) d.c. compound motor
 c) a.c. split-phase motor
 d) a.c. three-phase motor

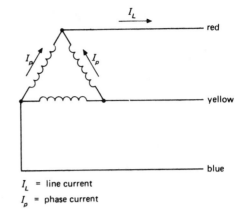

I_L = line current
I_p = phase current

Fig. 8.38

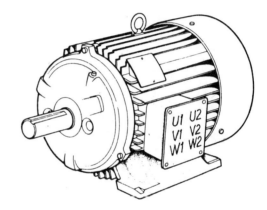

Fig. 8.39

Answers to exercises

Exercise 1

1. a) 350 mΩ
   b) 0.75 J
   c) 0.255 MW
   d) 400 kV
   e) 22 μF
2. a) η For comparing output/input in the same units
   b) π The constant for any circle
   c) ϱ A material's specific resistance
   d) ∝ Two quantities varying by the same amount
   e) ⩾ A quantity may either have the same value as another quantity or alternatively be a higher value
3. a) $d = \sqrt{(4A/\pi)}$
   b) $d = 1.78$ mm
4. a) 10^{11}
   b) 10^{-2}
5. a) $V_s = 250$ V
   b) $\cos \phi = 150/250 = 0.6$
   $\phi = 53°$
6. a) 6 cm^2
   b) right-angled triangle
7. $V = 300 \, \pi \, (3^2 - 2.4^2)/4$
   $= 763.5$ cm^3
8. Draw a straight line of best fit near the point plotted. Create a right-angled triangle on the slope of the line. The y-axis represents a small increment of voltage and the x-axis a small increment of current. Since $R = v/i$, then the approximate value of $R = $ is 2.35 Ω

9. On your sketched half sinewave, draw and measure six mid-ordinates. The maximum value of the sinewave is to be considered unity and therefore no mid-ordinate must exceed 1. For the average value of induced e.m.f.
$E_{ave} = (e_1 + e_2 + \ldots e_6)/6 = 0.637$
$E_{rms} = \sqrt{[(e_1^2 + e_2^2 + \ldots e_6^2)/6]} = 0.707$
Note: Answers will only be approximate.
10. The ratio $V_p/V_s = 20$
therefore $V_s = 6600/20 = 330$ V

Exercise 2

1. a) $R_e = 44$ Ω
   $P = V^2/R = (220 \times 220)/44 = 1100$ W
   b) $R_e = 4$ Ω
   $P = V^2/R = (220 \times 220)/4 = 12100$ W
2. a) In the parallel branch $C = 10$ μF
   For the whole circuit $C = 2$ μF
   b) $Q = CV = 2 \times 10^{-6} \times 100 = 200$ μC
   c) $W = \frac{1}{2}CV^2 = \frac{1}{2} \times 2 \times 10^{-6} \times 100 \times 100 = 0.01$ J or 10 mJ
3. a) $I = 280$ A
   b) Since voltage drop $V = IR$,
   Voltage drop at A is $V = 280 \times 2 \times 50 \times 0.0001 = 2.8$ V
   Voltage at A is $240 - 2.8 = 237.2$ V
   Voltage drop at B is $V = 130 \times 2 \times 40 \times 0.0001 = 1.04$ V
   Voltage at B is $237.2 - 1.04 = 236.16$ V
   Voltage drop at C is $V = 50 \times 2 \times 40 \times 0.0001 = 0.4$ V

Voltage at C is 236.16 − 0.4 = 235.76 V

Voltage drop at D is $V = 30 \times 2 \times 30 \times 0.001 = 0.18$ V

Voltage at D is 235.76 − 0.18 = 235.58 V

4 $R = 48\ \Omega$

Since $R = \varrho l/A$ amd $\varrho = RA/l$

$$\varrho = \frac{(48 \times 10 \times 10^{-6})}{10} = 48\ \mu\Omega m$$

5 a) $F = BIl$ newtons $= 25 \times 0.05 \times 10$
       $= 12.5$ N

   b) One example is in the operation of a *moving-coil instrument* and another is in the operation of a *motor*. The interaction has already been described on page 39 of this chapter. Figure 2.19 shows the distortion of the main field as a result of a conductor field. The stretched main field magnetic flux tries to straighten out and in doing so it ejects the conductor.

6 The difference between alternating current and direct current electricity is the way it is finally produced from the generator. An a.c. supply passes through the slip-rings whereas a d.c. supply passes through a commutator which inverts the negative cyclic action that would otherwise be created.

   A transformer relies on electromagnetic induction for its operation and an a.c. supply provides a continuing changing magnetic field for the purpose of creating an induced e.m.f. This e.m.f. can be made large or small depending upon the number of turns being cut by the magnetic field. See Chapter 5.

7 Since $\alpha = (R_t − R_0)/R_0 t$
   then $\alpha = (330 − 300)/(300 \times 25)$
          $= 0.044\ \Omega/\Omega/°C$

8 Since $e = −L(i_2 − i_1)/t$
   then $e = 0.7(10 − 2)/0.04$
          $= 140$ V

9 a) $I = 0.961$ A
       $V = 9.61$ V

   b) $I = 0.149$ A
       $V = 1.49$ V

   c) $I = 0.43$ A
       $V = 4.3$ V

10 A diagram for this circuit is shown in Figure 2.30.

Note: Most of the above answers are fully explained in *Worked Examples in Electrical Installation* by the same author.

Exercise 3

1 a) (i) $Z = 20\ \Omega$, (ii) 12 A, (iii) 144 V and 192 V

   b) See Figure 61, page 70 in *Worked Examples in Electrical Installation* (*WEEI*) by the same author.

2 See *WEEI* page 76

3 See *WEEI* page 77

4 See *WEEI* page 77

5 See *WEEI* page 78

6 See *WEEI* page 81

7 (a) 16.93 A, (b) 0.8 lagging, (c) 10.322 kW, (d) 12.9 kVA

8 See Figure 3.1 as to the shape of the graphs.

9 Follow procedure in Chapter 1, Terminology.

10 a) Reduces supply cable size and switchgear costs.

    b) Impossible to have a balanced system. The neutral provides the system with an alternative path allowing the phases to be used economically.

    c) See Figure 3.24 in chapter.

Exercise 4

1 See *WEEI*, page 65.

2 a) (ii) Since $P = VI \cos \phi$ then $VI = 240 \times 7$
                 $= 1680$ VA or 1.68 kVA

       (ii) $\cos \phi = P/VI = 1.1/1.68 = 0.655$ lagging

       (iii) Efficiency = Output/Input
              Output = Input × Eff. = 1100 × 0.72 = 0.792 kW

   b) See Section 552, *IEE Regulations*.

   c) (i) replacing a guard or carrying out maintenance work and supply failure where the motor was left switched on.

       (ii) in situations where the motor is vital such as an air compressor or life support machine.

3 a) The clearance between a motor's rotor/armature and its stator/main field poles.

b) The rotating part of a commutator motor.

c) The part of an armature on which rests brushgear in order to conduct current into the armature windings.

d) The induced, generated voltage which acts in opposition to the supply voltage.

e) The travelling magnetic field speed set up by the frequency of supply.

4 *a)* See page 74 of this chapter.

   b) (i) P = Output/Eff. = 4/0.8 = 5 kW

      (ii) $I = P/V\cos\phi$ = 5000/250 × 0.8 = 25A

     (iii) $Q = \sqrt{S^2 - P^2}$

$$= \sqrt{6.25^2 - 5^2}$$

$$= 3.75\text{kVAr}$$

5 See motor installations and Q4 Exercise 8 in 'Theory and Regulations' book by same author.

6 *a)* (i) Figure 4.27 shows sketches of these devices. In the magnetic device, the motor current passes through a solenoid which supports a soft iron armature and piston inside a pot of oil. The movement of the armature is controlled by the amount of overload current and if this is excessive it will cause the trip mechanisms to be activated.

      (ii) In the bimetallic overload device, any excess current taken by the motor windings will result in heat causing the device to bend and operate the trip contacts of the circuit. In practice both devices have selection on the amount of overload that can be allowed before their trip mechanisms operate.

   b) (i) Since P = Output/Eff. = 4000/0.76

$$= 5263 \text{ W } (5.26 \text{ kW})$$

      (ii) $I = P/V \cos\phi$ = 5263/(240 × 0.85)

$$= 25.8 \text{ A}$$

7 *a)* This is to overcome the high starting current the motor's armature will take when it is first switched on.

   b) See motor installations in 'Theory & Regulations' book by same author. The NVR acts as an electromagnet when the supply current is flowing and this holds the handle in the 'on' position. When the supply fails the electromagnet releases the handle as a result of spring action. The OLR is also an electromagnet connected in series with the armature circuit. An overload will operate the device and cause its armature to short circuit the NVR contacts, and the starter handle will again return to the off position.

8 See motor installations in 'Theory & Regulations' book by the same author.

9 For the no-load conditions

$I_f = V/R_f$ = 240/120 = 2 A

$I_a = I_L - I_f$ = 5 − 2 = 3 A

$E_b = V_s - I_a R_a$ = 240 − (3 × 0.2) = 239.4 V

For the load conditions

$I_a = I_L - I_f$ = 40 − 2 = 38 A

$E_b = V_s - I_a R_a$ = 240 − (38 × 0.2) = 232.4 V

Since $E \propto \Phi n$, then $E_1/E_2 = n_1/n_2$

239.4/232.4 = 30/n_2

therefore n_2 = (30 × 232.4)/239.4 = 29.12 rev/s

10 See Chapter 4 on electric motors.

Exercise 5

1 See Figure 5.34

2 *a)* 20:1

   b) 30 turns

   c) 30/12 = 2.5 volts/turn

   d) N = 4 × 2.5 = 10 turns

3 See Figure 5.35

4 *a)* See Figure 5.6. The core will become hot when the transformer is on load and oil convection will occur upwards and into the tubes which are designed to be in direct contact with the outside air. The oil will then return to the base of the transformer core at a much lower temperature.

   b) (i) Fall in oil level owing to a leak, (ii) winding fault.

5 See earthing arrangements in 'Theory & Regulations' book by same author.

6 *a)* T_1 is a thyristor being gated or triggered by the variable control resistor (R_2). Once triggered the thyristor will conduct and operate the motor.

   b) R_1 is a current-limiting resistor since the whole of this circuit is across the main supply.

   c) R_2 controls the pulses to the thyristor.

   d) D_1 blocks a.c. to the control circuit.

   e) D_2 conducts only positive-going pulses.

7 *a*) This is called a thyristor.
  b) See Chapter 5 on this device.
  c) See Figure 5.36.
  d) See Figure 5.36
8 *a*) See notes in the chapter.
  b) The withstand voltage of a rectifier without breaking down.
9 *a*) See page 63, Q109, *WEEI* book.
  b) See notes in this chapter.
10 *a/b*) See diagrams and notes in the chapter.
  c) Forward peak and reverse peak voltages.

Exercise 6

1 See page 94, Figure 96, in *WEEI* book.
2 See page 99, Figure 97, in Vol. 3, *EIT* Advanced Work book.
3 See procedure in chapter. Answer is 2 hrs 20 mins.

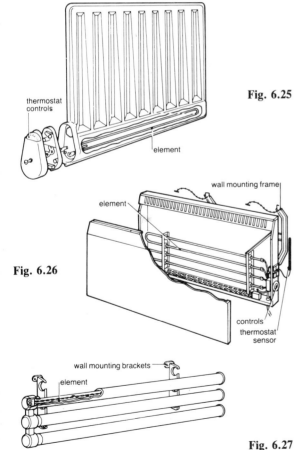

Fig. 6.25

Fig. 6.26

Fig. 6.27

4 (i) See Figure 6.25. The radiator uses a light-grade oil as its heat transfer medium. It operates by an immersion heater element which then transmits heat mostly by convection air currents. It finds a lot of use in passageways and under windows.
  (ii) See Figure 6.26. The heater often takes the shape of a rectangular metal panel and encloses a heating element. Air ventilation is provided at the top and bottom of the heater to encourage convection air currents. It has a general use.
  (iii) See Figure 6.27. The heater comprises steel or aluminium tubes enclosing an internal heating element. It radiates heat to the tube surface but most of its heat is transmitted by convection.

5 97%
6 At B the illuminance is 50 lx
   At C the illuminance is 17.67 lx
7 See page 69, Figure Q 123 in *WEEI* book
8 *a*) (i) The efficacy of a 400 W elliptical-shaped MBF lamp is approximately 3½ times that of a 500 W single-coil tungsten lamp. Also the MBF lamp's nominal life is the order of 7500 hours compared with 1000 hours for the tungsten lamp. What this really means is that the higher the luminous efficacy (lm/W) of the lamp installation the less maintenance cost incurred over a period of time, such as lamp replacement costs and labour costs.
    (ii) One operational disadvantge with the mercury discharge lamp is that if the supply to it fails for a brief moment, it will not relight immediately – it takes 2–7 min to restrike.
    (iii) A mercury vapour lamp of 400 W will require control gear for its operation and the ballast component for this lamp dissipates approximately 24 W. When dealing with discharge lamps in general, one has to add this loss to the lamp wattage rating before ascertaining the efficacy of the luminaire (lighting fitting).

b) The solution to this part of the question is shown in Figure 6.28.

$$\text{power factor } (\cos \phi) = \frac{P}{VI}$$

$$= \frac{440}{240 \times 3.5}$$

$$= 0.52 \text{ lagging}$$

thus $\cos^{-1} 0.523 \qquad = 58°23'$

By construction and measurement it will be seen that the capacitor improves the current taken by the supply to 2.3 A and its phase relationship with the supply voltage is approximately 38° lagging.

9 *a)* Firstly, one needs to find the power factor of the circuit before the switch is closed.

$$\text{power factor } (\cos \phi) = \frac{\text{power } (P)}{\text{voltamperes } (VI)}$$

$$= \frac{68}{240 \times 1.2}$$

$$= 0.236 \text{ lagging}$$

thus $\cos^{-1} 0.236 \qquad = 76°21'$

This is called the phase angle under these conditions.

Secondly, to construct a phasor diagram (i.e. a diagram showing current and voltage quantities rotating about a fixed point in an anticlockwise direction), choose a suitable scale for the current value (say 1 A = 5 cm) and with a protractor, mark out 6 cm, lagging the voltage reference line by 76° 21'. There is no need to draw the voltage line to any scale.

When the switch is closed, the capacitor circuit takes a leading 1 A and 5 cm needs to be marked out at 90° to the voltage reference line. The current taken by the supply when the switch is closed is found by constructing a parallelogram as shown in Figure 6.29. This is found to be 0.33 A and lags behind the supply voltage by a phase angle of approximately 30°. The capacitor improves the power factor of the circuit to 0.866 lagging.

Fig. 6.28

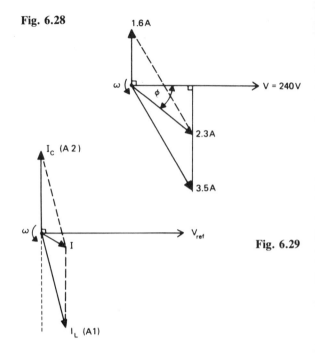

Fig. 6.29

b) The wattmeter reads 68 W, which is the total power dissipated in the circuit. In order to determine the efficacy of the lamp, the ballast losses need to be deducted from the wattmeter reading. Thus:

$$\text{efficacy} = \frac{\text{lumens}}{\text{watts}}$$

$$= \frac{5500}{55}$$

$$= 100 \text{ lm/W}$$

c) The resistor R across the capacitor is for discharge purposes. Unless the capacitor is fitted with a discharge resistor, it may hold a charge resulting in an electric shock. See *IEE Wiring Regulation 461–4*.

d) It is pointed out in the *Electrical Equipment (Safety) Regulations, 1975* that:

Where it is necessary for safe operation of any equipment to which these Regulations apply, that the user should be aware of any particular characteristic of equipment, the necessary information shall be given by markings on the equipment itself, or where this is not practicable, in a notice accompanying the equipment.

It is therefore most important to read the manufacturer's instructions and leaflets enclosed within the lamp packings of *all types* of lamp.

With regard to SOX and SLI lamps, they contain metallic sodium, and lamps for disposal must be kept dry to prevent water reacting with this element. The reaction is violent and there is a real risk of fire.

Disposal must include precautions against flying glass, breaking the inner tube and deactivating the metallic sodium with water under controlled conditions. If the outer bulb is broken for disposal, precautions must be taken against flying glass or other fragments. The operation should be carried out in a well-ventilated area (preferably outdoors) and in a proper container in a dry atmosphere.

10 See page 81, *WEEI* book.

Exercise 7

1 See *WEEI*, page 61
2 (i) 61.7 Ω, (ii) 328 Ω, (iii) 995 Ω
3 See *WEEI*, page 80
4 See *WEEI*, page 84
5 0.75 lagging
6 $I_L = 130$ A
7 Hints: Draw voltmeter connected to selector switch.
        Draw four supply lines to selector switch. Create selector switch internal circuit such that one voltmeter connection is permanently on the neutral and the other is in the centre of the switch and able to rotate and make contact with any phase.
8 See text in chapter
9 See Chapter 5 on current transformers
10 (i) 114 kW, (ii) p.f. = 0.978.

Answers to multiple-choice questions

Part I Certificate

1	*c*	11	*c*	21	*d*	31	*c*	41	*d*
2	*d*	12	*a*	22	*b*	32	*a*	42	*c*
3	*a*	13	*d*	23	*c*	33	*b*	43	*c*
4	*a*	14	*b*	24	*c*	34	*b*	44	*b*
5	*b*	15	*c*	25	*a*	35	*b*	45	*a*
6	*c*	16	*b*	26	*c*	36	*d*	46	*a*
7	*c*	17	*c*	27	*c*	37	*c*	47	*d*
8	*d*	18	*b*	28	*b*	38	*b*	48	*c*
9	*a*	19	*d*	29	*d*	39	*d*	49	*c*
10	*c*	20	*c*	30	*a*	40	*c*	50	*b*

Part II Certificate

1	*c*	11	*a*	21	*c*	31	*a*	41	*a*
2	*a*	12	*d*	22	*c*	32	*d*	42	*d*
3	*a*	13	*d*	23	*b*	33	*b*	43	*c*
4	*d*	14	*a*	24	*c*	34	*c*	44	*d*
5	*d*	15	*a*	25	*d*	35	*d*	45	*c*
6	*c*	16	*d*	26	*b*	36	*a*	46	*a*
7	*b*	17	*b*	27	*c*	37	*b*	47	*d*
8	*b*	18	*d*	28	*d*	38	*c*	48	*a*
9	*c*	19	*a*	29	*c*	39	*b*	49	*c*
10	*b*	20	*c*	30	*d*	40	*a*	50	*d*

Note: Some comments on these answers and the distractors are given in *Multiple-Choice Questions in Electrical Installation Work* by the same author.

Index